Advanced Fermentation and Cell Technology

Advanced Fermentation and Cell Technology

**Urvashi Swami and
Kunwar Digvijay Singh Thakur**

Delve
PUBLISHING
www.delvepublishing.com

Advanced Fermentation and Cell Technology

Urvashi Swami and Kunwar Digvijay Singh Thakur

Delve Publishing
224 Shoreacres Road
Burlington, ON L7L 2H2
Canada
www.delvepublishing.com
Email: orders@arclereducation.com

ISBN: 978-1-77469-100-7 (Hardcover)

ABOUT THE AUTHORS

Dr. Urvashi is a microbiologist. She completed her Masters from Chaudhary Charan Singh Haryana Agricultural University, Hisar, after which she joined Panjab University, Chandigarh for completing doctorate. Her main area of research was fermentation and phytotherapy. After completion of PhD, she moved to Kuwait where she is running her own business of medical writing and editing.

Kunwar Digvijay Singh Thakur is an enthusiatic learner, researcher and academic writer with formal degrees in biotechnology and laws. He has drafted and editorialized many scientific projects at college and university level. An amateur trekker, avid reader and chef at heart, he has a nerve for learning new pursuits and improvising with his academic skills.

TABLE OF CONTENTS

LIST OF FIGURES

LIST OF TABLES

LIST OF ABBREVIATIONS

5′-GMP	guanosine 5′-monophosphates
5′-IMP	inosine 5′-monophosphates
AA	acetic acid
ADH	alcohol dehydrogenase
AldDH	aldehyde dehydrogenase
ASS	adenylosuccinate synthetase
ATP	adenosine triphosphate
BCG	*Bacillus* Calmette-Guérin
BOD	biological oxygen demand
CAGR	compound annual growth rate
CBHs	cellobiohydrolases
CBM	carbohydrate-binding module
CEs	carbohydrate esterases
CGD	chronic granulomatous disorder
CHO	Chinese hamster ovary
CO_2	carbon dioxide
COD	chemical oxygen demand
CPT	camptothecin
DAPA	diaminopimelic acid
DNA	deoxyribonucleic acid
DSP	downstream processing
EMP	Embden-Meyerhof pathway
EPS	exopolysaccharide
ER	endoplasmic reticulum
FAD	flavin adenine dinucleotide
GA	glucoamylase
GHs	glycoside hydrolases
GI	gastrointestinal tract

GMP	good manufacturing practices
GO	glucose oxidase
GOS	galactooligosaccharides
HPLC	high-performance liquid chromatography
HPV	human papillomavirus
HSD	homoserine dehydrogenase
IDH	isocitrate dehydrogenase
IFN-α	α-interferon
IMP	inosine monophosphate
LA	lactic acid
LC-MS	liquid chromatography-mass spectrometry
LF	liquid fermentation
LOX	lipoxygenases
LPF	liquid phase fermentation
MAb	monoclonal antibodies
MPF	mixed-phase fermentation
MSM	microbial synthetic metabolites
NH_3	ammonia
NH_4^+	ammonium
O_2	oxygen
PDC	pyruvate decarboxylase
PLA1	phospholipase A1
PLA2	phospholipase A2
PLC	phospholipases C
PLD	phospholipases D
RER	rough endoplasmic reticulum
rRNA	ribosomal ribonucleic acid
SCID	severe combined immune deficiency
SCP	singled cell protein
SER	smooth endoplasmic reticulum
SMF	submerged fermentation
SPF	solid-phase fermentation
SSF	solid-state fermentation
SSI	site-specific immunotherapies

STD	sexually transmitted disease
TCA	tricarboxylic acid
TPC	total phenol content
VLPs	virus-like particles
WBC	white blood cells
WWT	wastewater treatment

PREFACE

Many traditional enzyme and microbial systems are included in fermentation and cell culture technologies that have been employed in the pharmaceutical, biochemical bioenergy, and agri-food industries. Novel technologies like protein engineering, systems biology, genetic engineering, and plant cell and mammalian cell systems, as well as the requirement for biomaterials, pharmaceuticals, bioenergy, and sustainable bio ingredients, are fast advancing. Industrial practitioners, students, and researchers must develop and implement new fermentation methods, and a range of cell technologies as the biobased economy continues to drive innovation.

Written by an expert in fermentation and cell technology, this comprehensive book contains information about the plant cells, mammalian cells, and various microbial technologies used in the contemporary biochemical processes. This convincing textbook illustrates the association between cell culture biopharmaceutical actives and food fermentation in conjunction with the crucial features of advanced fermentation and cell technology. Comprehensive chapters of this book have been divided into different that explain plant and animal cell technology, microbial cell technology, new biotechnologies' safety concerns, and the use of microbial fermentation in food items, pharmaceuticals, and chemicals.

This book is composed of eight chapters, and all these chapters explain a particular topic. To start with Chapter 1, readers will find a thorough introduction of fermentation and cell technology along with animal and plant cell cultures. Chapter 2 gives a detailed explanation of microorganisms in the fermentation process with their methods and diversity.

In Chapter 3, readers will be familiarized with the concept of fermentation miniaturization. Chapter 4 discusses the concept of aerobic and anaerobic solid-phase fermentation (SPF) with their advantages and disadvantages, phase control, and bioreactors.

Chapter 5 offers the explanation of the scaling-up of the industrial microbial process with its procedures. Chapter 6 thoroughly discusses the dietary uses of microbial fermentation along with a detailed explanation of several enzymes. Chapter 7 covers food quality and explains the role of microbial fermentation in food quality enhancement. This book ends with Chapter 8 that explains the medical applications of industrial fermentation, its various types, stages, and products.

Advanced Fermentation and Cell Technology is an ideal reference for technologists, researchers, and food scientists through the food industry, mainly in the fermented beverage, bakery, and dairy sectors, and is a valuable source for students of microbiology, biotechnology, food science, agricultural sciences, biochemistry, and biochemical engineering.

Chapter 1

Introduction to Fermentation and Cell Technology

CONTENTS

1.1. INTRODUCTION

To produce enzymes, biomass, pharmaceuticals, and chemicals, the processes of fermentation may be utilized. In such culture processes, traditionally used cell types are fungi, bacteria, and yeasts. Withal, lately, industrial fermentation processes incorporating plant and animal cells are being introduced (Marino and Gaggìa, 2013). Reactor configurations are required by the aerobic fermentation incorporating microbial cells which expedite oxygen transfer and aeration. These include rotating-drum, air-lift reactors, stirred-tank reactors, and tray systems concerning solid substrate fermentation, and several other specialized systems. A simpler bioreactor design is required by the anaerobic fermentation as there is no need for aerated facilities. Minerals, nitrogen, carbon, and at times growth factors, oxygen, and water (if aerobic) are required by the microorganisms as energy sources for cell maintenance and biosynthesis, and as constituents for cell biomass. Growth factors are not required by some of the microorganisms in the medium, whereas others need complex media involving some nucleotides, vitamins, or amino acids. The design of the media and the parameters of alternate fermentation is in such a way that optimizes both product formation and cell growth. Chemical and physical environmental conditions affect the yield and cell growth rate, involving pH, temperature, toxic metabolites production, and the presence of growth-limiting substrates. Medium constituents (e.g., inducers, precursors, inhibitors, repressors), morphology, growth rate, temperature, oxygen/carbon dioxide (CO_2) concentration, rate of substrate assimilation, pH, and secretion of by-product and other factors may affect the formation of product. They need conditions of gentler agitation in culture and are very fragile as mammalian cells are inadequate to a tough outer cell wall (against microbial cells).

Moreover, their nutritional needs are too complicated, and cells are very receptive to change in parameters like pH, temperature, CO_2, and dissolved oxygen. Based upon cell type, animal cells can grow in monolayers cultures connected to surfaces (i.e., anchorage-dependent) or being suspended cells (i.e., anchorage-independent). The contact inhibition property is shown by some animal cells, that is, they quit raising on the contact surface with other cells (Fröhlich, 2018). These properties dramatically affect environmental conditions and the nature of media used in the culture of an animal cell. Sometimes, air-lift or gently stirred fermenters concerning the growth of suspension cultures are involved in the reactor types, which incorporates the usage of microcarrier beads concerning anchorage-dependent cells. For the elimination of sensitivity of cells to shear, cell encapsulation has been utilized. Semi-permeable membranes, in other systems, are used to regulate concentrations of nutrients, toxic metabolites, and gases through perfusion. Usually, newborn calf-, fetal calf-, calf-, or horse-serum (5–10% v/v) involved in the animal cell culture media creates a supply of growth factors. Where for a specific cell type, there is an optimization of basal media, where the serum content can be minimized to 1 to 2%. Usually, these basal media involve 13 to 22 amino acids, 6 to 12 types of inorganic salt, glucose, 9 to 13 vitamins, and alternate compounds. The detection and supply of the key serum-containing growth factors are required by the formulation and development of serum-free media for usage in mammalian cell culture (Jayme and Blackman, 1985). In suspension culture, there is a growth of several plant cells in chemically defined media involving a nitrogen source like nitrate, and organic carbon source like sucrose, growth regulators, and inorganic salts (Mir et al., 2018).

1.2. MICROBIAL FERMENTATION PROCESSES EXAMPLES

1.2.1. Alcohol Production

Perhaps, the most significant example of anaerobic fermentation is the formation of alcohol using fermentable sugar. The oxidative enzymes of the cytochrome system and the tricarboxylic acid (TCA) cycle are repressed catabolite at higher than 250 mg/l glucose concentrations, and yields of yeast biomass remain little with many sugars being transformed to alcohol under aerobic conditions. Alcohol yields are increased further by the maintenance of anaerobic conditions and minimized biomass production because of the phenomenon called the Pasteur effect (Gualdrón et al., 2012). Efficient recycling of minimized NADH under such conditions, formed at the glyceraldehyde-3-phosphate dehydrogenase phase in glycolysis is concluded from the ensuing reduction of acetaldehyde to ethanol. The 0.49 g CO and 0.51 g ethanol are the theoretical yields by 1 g glucose. Practically, the glucose of around 10% is transformed to biomass. Therefore, yields of CO and ethanol, may attain theoretical values of 90%. Whereas alcoholic fermentations are highly anaerobic, a little oxygen is required to facilitate the yeast to unsaturated fatty acid membrane elements and synthesize little sterols. For the supplementation of the fermentation medium with oleanolic or oleic acid, this need for oxygen vanishes.

Around 12% to 14% of ethanol concentrations can be achieved by several yeast strains. However, choosing strains has been shown to be able to produce alcohol of 18 to 20%, the rate of fermentation is reduced greatly with the increase of ethanol concentration (Contreras et al., 2014). The yeast's alcohol tolerance is affected greatly by the phospholipid composition of the yeast plasma membrane. Increased alcohol tolerance is observed when membrane unsaturated fatty acid (incomplete sentence).

Figure 1.1(a) illustrates the time course of a conventional batch molasses fermentation. The density of the initial yeast cell comes to about 5 to 10 billion cells per liter; it increases to around 100 billion at the completion of fermentation (Collett, 1851). The primary growth stage ends at around 15 hours, with the accumulative fermentation consuming approximately 30 hours. Around 80% cells are recycled in the process of Melle-Boinot (Logomasino, 1949), thus, the density of the initial yeast becomes 80 billion cells per liter, consequent to a greater rate of the production of the ethanol with a too minimized time of fermentation. Figure 1.1(b and c) illustrates the instance patterns of alcohol production and yeast growth patterns in methods for the production of white wine and Bourbon, respectively. It must be noted that the lower levels of the yeast biomass turn to protracted fermentation times. Generally, variations in the alcohol fermentation rate in a broad range of processes can be associated with the variables like the fermentation pH, the fermentation composition of broth nutrients, and the yeast concentration and temperature.

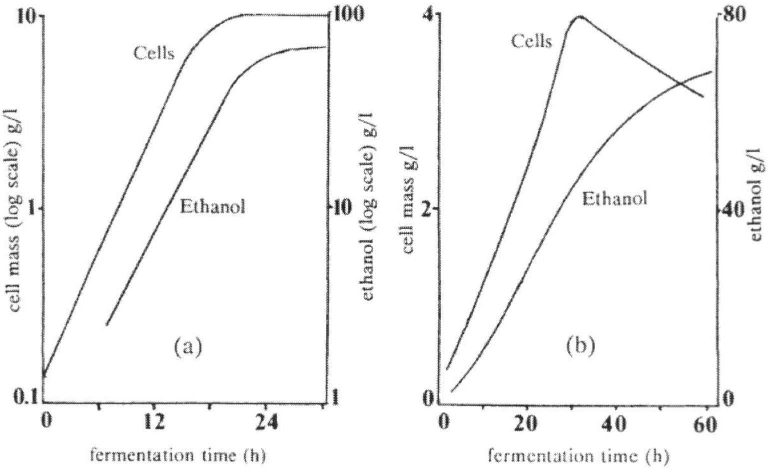

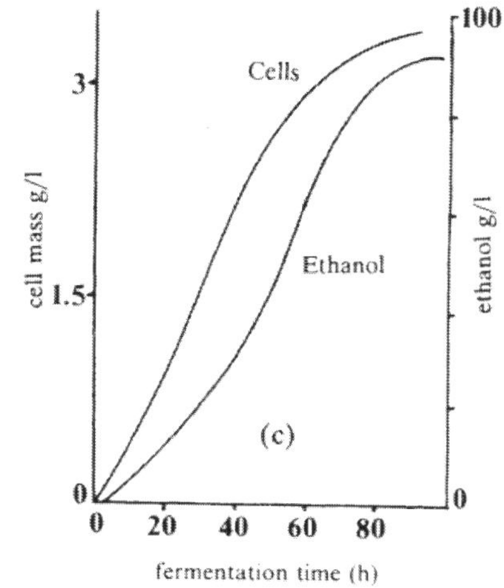

Figure 1.1. Ethanol production example processes. (a) Conventional batch molasses fermentation; (b) bourbon fermentation; (c) white wine fermentation.

Source: https://link.springer.com/chapter/10.1007/978-1-4615-3914-8_8.

1.2.2. Production of SCP

For the single-cell protein production from Fusarium graminearum, in the process of fermentation, for human consumption (refers to Figure 1.2), the elements of the medium involve gaseous ammonia, food-grade glucose, biotin, and salts. At 6.0, the fermentation pH is regulated from the inclusion of gaseous ammonia that is given from the air inlet stream. Up to 0.2 h^{-1} of the particular growth rate is obtained, and 15–20 kg m^{-3} of cell concentrations are produced (Solomons, 1983).

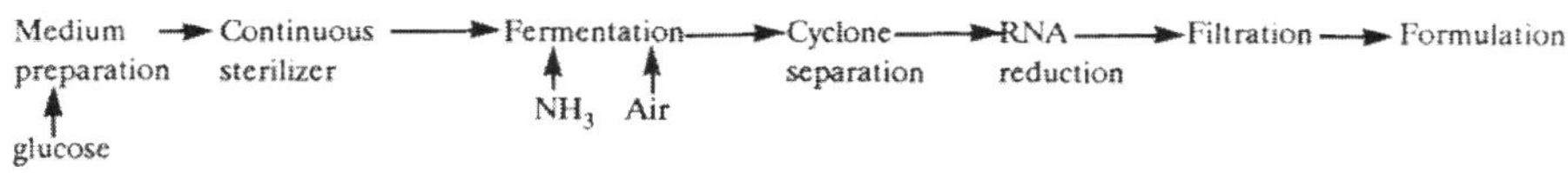

Figure 1.2. Fusarium graminearum produces SCP.

Source: https://link.springer.com/chapter/10.1007/978-1-4615–3914-8_8

In fermentations of aerated yeast, a low concentration of sugar is maintained by the constant sugar inputting facilitating biomass production. On the other hand, at aerobic or anaerobic conditions, a significant quantity of alcohol is formed for a high concentration of sugar. Urea or ammonium salts, molasses, biotin, and phosphate are the significant medium elements for the baker's yeast production (Junker and Wang, 2006). Up to 8 scale-up phases are involved in the batch fermentation process resulting in an enormous production fermenter of about 50 to 350 m^3. The starting two inoculum phases are aseptic, whereas the production stage and the next inoculum development stages are not and, hence, pressure vessels are not required. At 28 to 30°C, the exothermic fermentation temperature is upheld by cooling. The pH is upheld at 4 to 4.4 during the first half of the process of the production fermentation to minimize bacterial contamination. On the other side, a greater range of pH (i.e., 4.8 to 5) is utilized to minimize the molasses adsorption coloring matter on the yeast. There is a production of the yeast concentrations of 40 to 60 gl^{-1} (dry) (Rajarathnam et al., 1989).

1.2.3. Citric Acid

The requirement for the key fermentation for the formation of citric acid using the submerged cultivation regarding Aspergillus niger is the usage of a strong sugar concentration, normally in molasses, the inclusion of NIr4 during the fermentation, in the culture, the maintenance of manganese of low level, and immense aeration (Kubicek and Rohr, 1986). The pH fermentation is kept within five to seven for growth and A. niger germination, and after that, it is permitted to decrease below two for the production of citric acid. Lack of manganese has an influence on the reduction of the activity of few enzymes of the pentose phosphate pathway and TCA cycle in a way that offers increased production of citric acid. Lack of manganese impairs anabolism as well and stimulates the acid protease production, both are the reason for a significant raise in NH^+, concentration having rectification of an inhibitory influence of citrate upon glycolytic enzyme phosphofructokinase (Sanchez and Demain, 2008). Pyruvate carboxylase, an important anaplerotic enzyme, boosts citrate synthesis, also high activity of glycolytic enzymes is induced by a high concentrated carbohydrate. A significant respiratory pathway that is the SHAM-sensitive pathway is maintained by the high oxygen tension, which does reoxidization of NADH, although, not with normal respiratory chain ATP production. Transformation of glucose to gluconic acid using the inactivated glucose oxidase (GO) is prevented by the low culture pH.

In 1977, the initiation of a highly productive fermentation process for the formation of citric acid was done by the Candida guilliermondii. Table 1.1 illustrates the key features of this method. A high-capacity water cooling tower is required by the production plant, and there is an extremely exothermic fermentation. Constant bowl centrifugation obtains the disengagement of the Candida yeast cells by the citrate-involving fermentation liquor.

Table 1.1. Outline of a Key Facet of the Citric Acid Production by Candida guilliermondii

Aspect	Description
Fermentation	In air-lift bioreactors or large stirred-tank; pH: for growth 4.50–6.50, for citrate production, around pH 3.50. Temperature: 25.0–37.0°C; Aeration: 0.50–1.0 vvm
Characteristics of fermentation	Exothermic, high cooling water requirement, high fermentation rate
Medium	Glucose syrup or molasses (up to 280 kg/m^3) also additional salts and nutrients. No need for metal ion pretreatment. Usage of high sugar concentrations is allowed by the osmotolerant nature of the organism.
Requirements for acid production	Acid accumulation is triggered by the limitation of nitrogen. The requirement of thiamine for citrate accumulation.

1.2.4. Amino Acids

For the industrial formulation of the lysine and glutamic acid using fermentation, the strains of the bacteria Brevibaclerium and Corynebacterium are utilized (Vallino and Stephanopoulos, 2000). Due to cell metabolism, wild-type isolates hardly secrete specifically end-product repression plus inhibition, regulatory techniques, substantial stages of amino acids. Because of the fruitful progress of deregulated mutants, there is formation of a commercial number of amino acids (Nakayama, 1982). For instance, the initial enzyme is controlled by concerted response inhibition by L-lysine and L-threonine in the biosynthesis of lysine supported by B. flavum and C. glutamicum (refers to Figure 1.3). Production of good lysine is promoted by the cutback in the intracellular pool of threonine using the generation of threonine-methionine double auxotroph, or a homoserine auxotroph. There is a predomination of the Embden-Meyerhof-Parnas pathway in the formation of glutamic acid using hexose (Hirose et al., 1985).

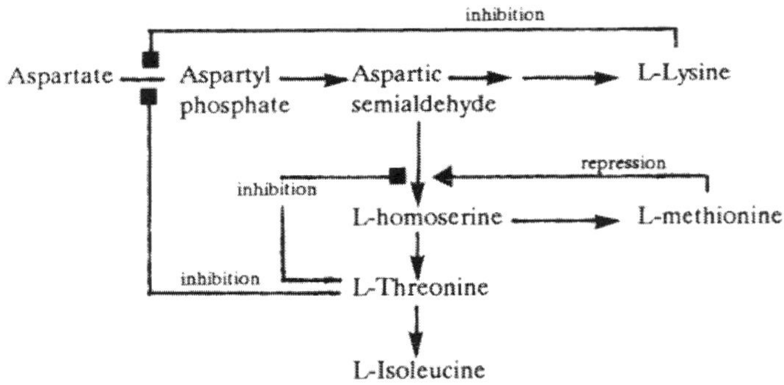

Figure 1.3. The control of L-lysine synthesis in B. flavum and C. glutamicum.

Source: https://link.springer.com/book/10.1007/978-1-4615-3914-8.

TCA cycle's isocitrate is transformed to a-ketoglutarate, which consequently is aminated reductively to produce glutamate. Replenishment of the cycle is required by the reduction of isocitrate using the TCA cycle through the anaplerotic methods, including the function of the glyoxylate cycle and the production of oxaloacetate by phosphoenolpyruvate. NH_3 is provided by the gaseous ammonia, which is inputted on a chronic basis for the amination phase and also supports pH regulation (Folbergrova et al., 1974). Sub-optimal stages of biotin carry out the fermentation. The accumulation of amino acid is controlled primarily using its excretion rate, and sub-optimal concentrations of biotin outcome in a declined concentration of phospholipids inside the cell membrane, raising cell permeability towards glutamate. A cofactor of acetyl-CoA carboxylase is biotin which is an initial enzyme in the biosynthesis pathway of oleic acid for inclusion with phospholipids (Vick and Beevers, 1978).

1.2.5. Penicillin

For the penicillin industrial production, Penicillium chrysogenum strains, using typical genetics incorporating selection and mutation, protoplast fusion, and parasexual breeding methods are used (Queener and Swartz, 1979). A range of acyl substituents are carried by the 6-aminopenicillanic acid (6-APA) fundamental structure, e.g., phenoxyacetic acid and phenylacetic acid in the cases of penicillin V and G, respectively. Glucose catabolite-repress the penicillin biosynthesis which is also respond-regulated by penicillin and lysine. Figure 1.4 illustrates the pattern of penicillin formation. Medium constituents involve ammonia, corn steep, side-chain precursor, salts, and a carbon source like molasses, lactose, or glucose within a process of fed-batch culture. Inputting ammonia aids respiration, avoids mycelia lysis and provides nitrogen in consideration of 6-APA synthesis, and constant carbon input avoids catabolite repression. The side-chain precursor is continually fed as well, and the pH is regulated at 6.50. In the initial two days of fermentation, there is a predomination of biomass growth, and subsequently, with time, penicillin production grows with the decline in growth rate.

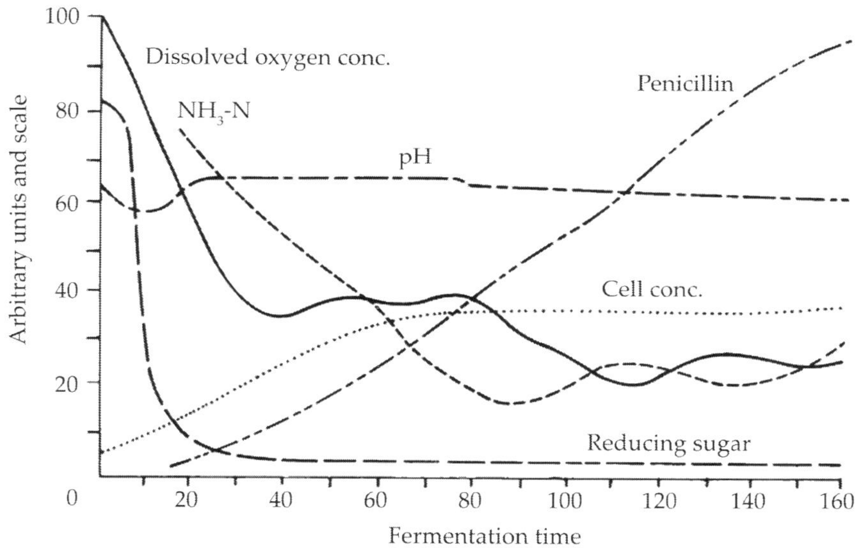

Figure 1.4. Classical sketch of the fermentation of fed-batch penicillin (Ryu and Hospodka, 1980).

Source: https://onlinelibrary.wiley.com/doi/abs/10.1002/bit.260220204.

1.2.6. Industrial Enzymes

Some of the significant features included in the industrial formation of extracellular enzymes are illustrated by the fermentation profiles regarding α-amylase by Bacillus amyloliquefaciens and submerged culture formation of glucoamylase (GA) by Aspergillus niger (refers to Figure 1.5). Classical fermentation media involve corn starch or ground corn being the carbon source, inorganic salts, and corn steep powder. The basic starch concentration in the fermenter is generally 50–100 gl^{-1}, liquefied partly through bacterial α-amylase in the temperature build-up preceding sterilization, which minimizes gelatinized starch viscosity (Keim, 1983). All along every process, more liquefied starch is included in fed-batch mode. Starch feeding in the Bacillus fermentation aims to keep the glucose concentration less than a critical repressive level. Glucose does not represent the GA formation in A. niger, and the principal goal of removing medium starch concentration using starch feeding is to minimize the viscosity of the medium. The pH reduces, in the A. niger fermentation, from around 6 to 4 all along the first 24 hours during the growth of active mycelial, and for many of the enzyme formations, the uncontrolled pH is within 3.50 to 4.0. With the inclusion of alkali within the process of bacterial α-amylase production, pH is regulated above 6.0.

1.2.7. Fermentations Along Recombinant Microorganisms

The usage of recombinant organisms is involved in the fermentation processes that are typically the same as the processes incorporating mutant or wild-type strains. No recombinant or natural organism has all of the features required for an effective fermentation process.

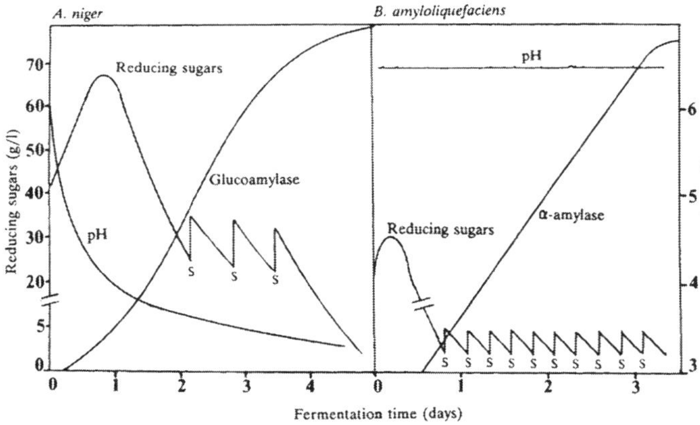

Figure 1.5. Fermentation sketches for the formation of glucoamylase using α-amylase and Aspergillus niger by Bacillus amyloliquefaciens [S: inclusion of starch].

Source: https://link.springer.com/chapter/10.1007/978-1-4615-3914-8_8.

However, these properties should be considered while making a recombinant strain and in progressing the process of fermentation (Taherzadeh and Karimi, 2007). These characteristics may involve: high promoter control, foreign gene stability, efficient product secretion, high product expression, rapid growth capacity, efficient glycosylation (if needed), high raw material consumption, high productivity and product yield, deficiency of the production of endotoxin, and reduction of recombinant product proteolysis.

A regulatable promoter, in several recombinant organisms, is involved in the expression vector. More processed control of the process of fermentation is facilitated, which usually enables induction to be turned on at a preferred time by the inducer inclusion to the medium. Growth might be retarded by the maximum rates of expression all along with the growth phase of the organism. Moreover, there might be a toxic influence of the high intracellular inflation of the products of recombinant foreign protein on cellular physiology, which makes regulatable promoters significant. The tryptophan (trp) promoter and the lac promoter are examples of significant controllable promoters utilized in E. coli expression vectors. For the direction of overproduction regarding foreign proteins, in E. coli, to a level of about 5 to 20% of soluble protein, the lac promoters are used. In some situations, many trp or lac promoters, or amalgam promoters that are derived from trp and lac, are found to raise expression (Atkinson et al., 1986). Subsequent to two tandem lac promoters, the entrance of a trp promoter inside a plasmid pHGH107 has led to the formation of an E. coli recombinant strain (RV308) that is able to produce 240 to 320 mg L^{-1} Met-hGH within a 500-L fermenter. Figure 1.6 illustrates the sort of cell growth and formation of product within a laboratory fermenter.

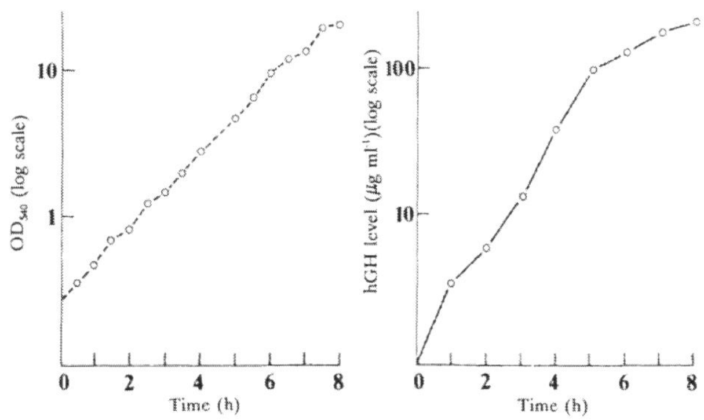

Figure 1.6. Culture attributes of a ten-liter growth culture of E. coli RV308 forming hGH in the existence of ampicillin (Atkinson et al., 1986).

Source: https://catalog.princeton.edu/catalog/396442.

It is shown that the temperature inducible PL promoter by phage can direct synthesis of 30 to 40% of the sum of the cellular protein of IRP synthetase A and cloned fi-lactamase (Remaut et al., 1981). Instances of expression vectors utilized in filamentous fungi like Aspergillus contain the A. nidulans alcohol dehydrogenase (alcA) promoter, the A. oryzae α-amylase promoter, and A. niger glucoamylase (glaA) promoter (Berka and Barnett, 1989). There are too efficient promoters coding, in yeast, for glycolytic enzymes, and hence, are normally used for the formation of expression vectors. The goal of a lot of techniques employed to express foreign genes within yeasts is the attainment of the high mRNA levels to increase gene expression (Martin and Scheinbach, 1989) by the usage of powerful yeast promoters and multiple copy plasmids, both regulated and constitutive. The extremely broadly utilized constitutive promoters used are derived from genes encoding for ADH1, G3PDH, PGK, and enolase. Regulatable promoters employed involve promoters that are incorporating to galactose metabolism, the promoter ADH2 and temperature regulated promoters, and different amalgam. A leading drawback of the heterologous proteins' expression within yeast is that there may be a toxic accumulation of foreign protein towards the cell. The physiology and nature of the organism and the physicochemical characteristics of the recombinant molecule influence the capability of organisms to release recombinant proteins, among other things. Industrial strains, for example, Aspergillus, and Bacillus subtilis species are known to be the immense extracellular proteins producers, and there is somewhat a partial understanding of the procedure of secretion involving the significance of secretion peptide leader sequences. Whereas normally, yeasts, and E. coli are feeble protein secretors, the inclusion of peptide secretion vectors within the construction of DNA has permitted the release of recombinant proteins to occur. For yeasts, the cell wall seems to establish a substantial barrier towards protein secretion. Protein, for example, invertase, which is released from the plasma membrane, is trapped in the periplasmic space within the cell wall and the plasma membrane. For the creation of human glycosylated recombinant proteins, there may be benefits of yeasts and filamentous fungi over bacteria because they can make glycoproteins, but the bacteria do not. Nevertheless, there are differences in the glycosylation patterns of yeasts and fungi in comparison with mammalian cells, which may influence the biological activity and proteins' tertiary structure (Werner et al., 2007).

1.3. ANIMAL CELL CULTURE

A growth curve is followed by the pattern of animal cells growth in a culture which is similar to that noticed with bacteria. However, the sluggish rate of mammalian cells growth, with strong well-adapted cells with a classical doubling time of 18 to 24 hours in contrast with below 60 minutes for several bacteria, is one of the factors that discerns the growth patterns. There is an increase in cell density, in a classical mammalian cell batch culture, which is from a starting cell inoculum of 10^5 cells/ml to an ultimate cell yield of 1–2 x 10^6 cells/ml. In vivo, cell concentrations can approach 10^9 cells/ml, which shows that the cells in a homeostatically balanced habitat determined by the animal's vascular system are so much evolved having high densities as compared to those noticed in batch culture (Chuong et al., 2002). The result of the attempts to make the cell yields rise within vitro through mimicking this perfusion system is the formation of cell densities larger than 10^7 cells/ml. The variety of bioprocessing strategies that have been employed for the formation of products in mammalian cell culture systems are illustrated in the following examples. Many examples have corresponded to techniques for the formation of monoclonal antibodies (MAb), which manifests the growing demand for such methods as there is an introduction of creative hybridoma methodologies (Taylor et al., 2013).

1.3.1. Viral Vaccines

Normally, in vitro animal-cell culture achieves the formation of viral vaccines regarding the human application, either killed or live. Live poliovirus is a combination of 3 kinds of attenuated polioviruses which have been spread in the cultures of human diploid cell lines or monkey kidney cells. In suspension culture, the mammalian cells are developed on micro-carrier beads and tainted with the poliovirus at the correct time. The virus grows quickly in the cells, resulting in cell lysis with the secretion of the virus within the medium (Butler, 1987).

1.3.2. Formation of Interferon by Lymphoblastoid Cells

For the suitable formation of interferon, a cell line is entrenched from a patient enduring Burkitt's lymphoma. The concluding B-type lymphoblastoid cell line is kept a near-diploid chromosomal complement having the ability to increase in suspension cell culture and continuous cell proliferation (Lazar, 1983). The culture is caused to make an α-interferon (IFN-α) after around 10 hours of incubation using the Sendai virus. Within the culture medium, there is secretion of interferon (109.0 units per liter) and can be withdrawn after inducing virus inactivation. The operation of this suspension culture process is commercially on a 10,000-liter scale to generate gram amount of interferon, and the cells are shown to be flexible to the formulations of low-cost serum-free medium which minimizes the issues of downstream processing (DSP).

1.3.3. Monoclonal Antibodies (MAb)

1.3.3.1. Manufacturing in Mouse Ascites Fluid

The initial typical process for the production of a monoclonal antibody contains the hybridoma cells injection into the peritoneal cavity of histocompatible mice, which leads to the production of ascites. Ascites is referred to a transudate which constitutes while responding to the agents of ascitogenic. A majority of hybridoma cells are referred to ascitogenic, and the MAb that are secreted are contained by the ascites that produce all along with the growth of hybridoma cells (Véliz, 2002). For the production of MAb, a classical method contains preintraperitoneal injection along with the alkane derivative pristane, which is a priming agent which increases the ascites yields. Every mouse is infused with 5×10^5 hybridoma cells around after 21 days, and using a sequential harvesting method, the antibodies are restored, it is a method in which the ascites fluid is 'tapped' 3 times in the period of 14 to 25 days after injection. This technique is devised and based on the observation, i.e., sequential tapping offers the maximum concentrations of MAb and the maximum yields of ascites from several mice present. The mean production volume of ascites is 4.80 ml per mouse, and the mean MAb concentration within ascites is 6.90 mg ml^{-1} (Chandler, 1987).

1.3.3.2. Production Through Cell Culture Using Perfusion and Batch Techniques

Some of the various approaches that are being employed to produce MAbs by cell culture are illustrated by the examples given below.

Contrast has been given between the rate of the production of antibodies in perfusion culture and stirred suspension (Seaver, 1987). In perfusion systems, the membranes employed are microporous (0.650 pm), openly penetrable into every molecule, involving the monoclonal antibody, however, maintaining the hybridoma cells. An accumulative of 107 cells has been added to the system's cell module that has primed with a medium of 300 ml and, finally, 5,700 ml medium is loaded. On the 8th day, and after that every 2 to 4 days, medium (1 liter) has been harvested and interchanged with fresh medium (1 liter). There has been a comparison of all antibodies made in this system with those made in constantly stirred culture. There is about twice the production of an antibody having cell line as immense in the perfusion culture in contrast with the continuing stirred culture (Figure 1.7). There is 5.80 ttg ml^{-1} average antibody concentration in the stirred system, while in the perfusion system, it is 1 I $\mu g\ ml^{-1}$.

A scientist describes a method outline regarding monoclonal antibody batch production using large-scale suspension cultures Lebherz (1987). There is a doubling time of the hybridoma, which created a murine IgG_2, i.e., 9 to 15 hours, and while exponential growth, uphold 90%–95% of viability. Dulbecco's Modified Eagle Medium is the main element of the production medium having 4.50 glucose, additionally fetal bovine serum, 7% (v/v) with petty supplements. In the 50-L New Brunswick tissue culture fermenter, working parameters are agitation rate, 100 rpm; pressure, 5 + 3 psi; temperature, 37°C; dissolved oxygen, higher than 10% saturation; pH, 7.0. Six passage phases are involved by the hybridoma culture scale-up from seed stock towards the 50-L fermenter.

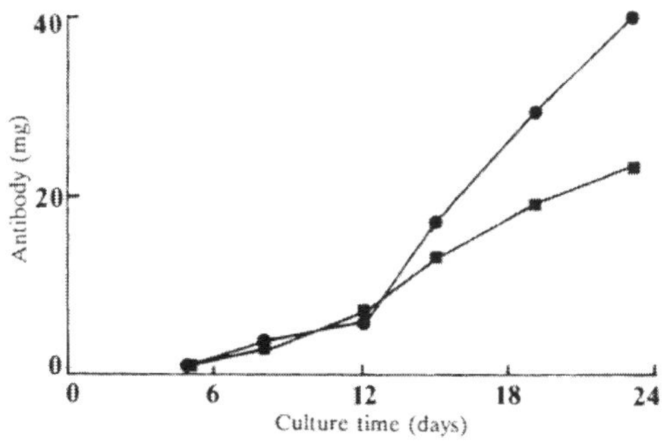

Figure 1.7. The contrast of accumulative antibody formation (mg) in perfusion (•) and constant stirred (■) culture by the line of Sp 2/0 hybridoma (Seaver, 1987).

Source: https://www.ncbi.nlm.nih.gov/pmc/articles/PMC1454779/.

Almost 5 to 7 days are taken by the cell growth for saturation following inoculation of the fermenters production with the production of monoclonal antibody growing to an utmost level in more 48 to 72 hours. The mean MAb concentrations in the supernatant is 72 pg. The preparation of monoclonal antibody is reduced by tangential flow filtration, purified using binding and elution to a protein A-Sepharose affinity column pursued by sterile filtration and 2nd concentration stage (Figure 1.8).

Figure 1.9 illustrates the production of antibody and cell growth patterns in microencapsulated cells (Possilico, 1986). Microcapsules of one volume, classically, is suspended in a medium of four volumes. Using a continued medium feed, microcapsule cultures are worked upon, having medium-throughput based on cell density.

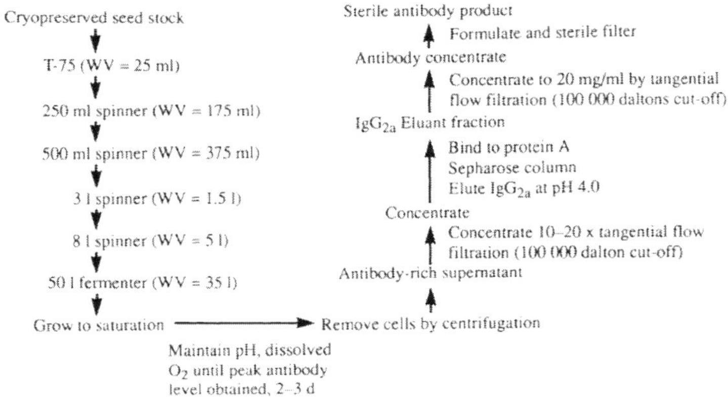

Figure 1.8. Method for massive formation of IgG, monoclonal antibody (re-formed by the consent of Lebherz, 1987).

Source: https://link.springer.com/book/10.1007/978-1-4615-3914-8.

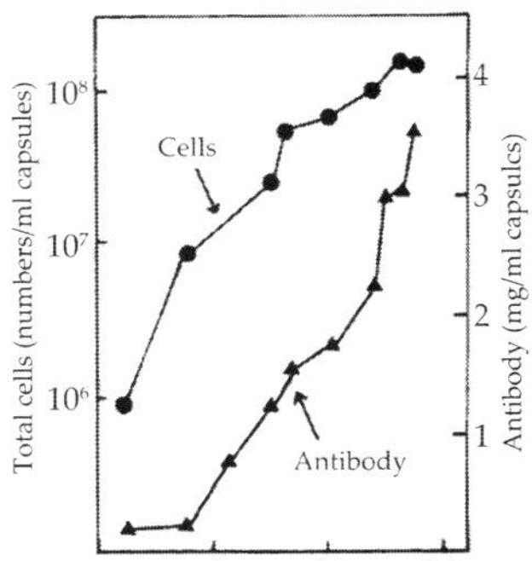

Figure 1.9. Antibody production and cell growth in microencapsulated cells (reformed by the consent of Posillico, 1986).

Source: https://link.springer.com/chapter/10.1007/978-1-4615-3914-8_8.

In microcapsules, accumulative medium volume employed for aiding cell growth is relatable to that needed for cell suspension cultures. One or more bioreactor core hollow fiber units, pumps, and oxygenators devised to allow efficient delivery of gases and nutrients are consisted by the techniques of mass culturing utilizing hollow fibers having continuous perfusion regarding production of monoclonal antibody. Figure 1.10 shows the indelible monoclonal antibody production by a murine hybridoma cell line kept in continuous perfusion for the period of five months (Lei et al., 2021).

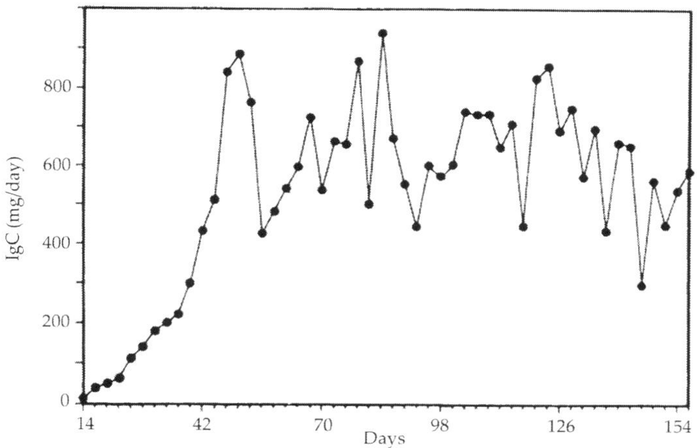

Figure 1.10. Durable constant perfusion culture of murine hybridoma excreting IgG in a chemically specified medium (reformed by the permission of Von Wedel, 1987).

Source: https://link.springer.com/chapter/10.1007/978-1-4615-3914-8_8.

An automated system, the Opticell culture system, is constructed for the animal cells mass production using ceramic matrices as growth chambers. Being a growth chamber, a cylindrical matrix having a length of 30.6 cm, a diameter of 9.1 cm, having a channel size of around 1.2 mm^2, is used. The whole matrix is involved within a flow loop, and the medium is continuously perfused all over the core and at the immobilized culture. The system is inoculated with 109 cells and filled with a medium of 20 to 40 L which ultimately evolve into immobilized cells in the matrix. In this 40 L perfusion system, to change the medium, two approaches can be used: a batch method by which the 40 L is exchanged after every 4 days, or a continual feed and medium harvest at a mean replacement of 10 to 12 L per day (Wurm, 2004).

When compared between the total IgG formed by the two techniques, it is found that the continuous method is more than 3 times efficient as compared to the batch feeding system.

1.4. PLANT CELL CULTURE

In Japan, the plant named Lithospermum erythrorhizon is conventionally used as a medicine because of its anti-inflammatory and antibacterial effects. The incorporated biologically active chemicals are the Shikonin and their derivatives, that are existed in the plant roots. There are also the applications of the Shikonin, which is a bright red naphthoquinone pigment, also used as a dye in cosmetics. There is a formation of a cell line that is able of gathering exceeding 15% (w/w) of biomass being a product. Medium elements involve indoleacetic acid, inorganic salts, and sucrose. The shikonin yield is 2.3 per gram, and the final cell yield is 17.5 per gram by following a fermentation production continuing for 14 days (Fujita and Hara, 1985).

REFERENCES

1. Atkinson, T., Barstow, D. A., Minton, N. P., Sharp, R. J., & Sherwood, R., (1986). High-level microbial expression and purification of recombinant proteins. Bioactive Microbial Products, 2, 27–43.
2. Benda, I., & Reed, G., (1982). Prescott and Dunn's Industrial Microbiology (Vol.1, pp. 19-39).
3. Berka, R. M., & Barnett, C. C., (1989). The development of gene expression systems for filamentous fungi. Biotechnology Advances, 7(2), 127–154.
4. Birch, G. G., & Shamil, S., (1988). Structure, sweetness and solution properties of small carbohydrate molecules. Journal of the Chemical Society, Faraday Transactions 1: Physical Chemistry in Condensed Phases, 84(8), 2635–2640.
5. Casa-Villegas, M., Marín-Navarro, J., & Polaina, J., (2017). Synergies in coupled hydrolysis and fermentation of cellulose using a Trichoderma reesei enzyme preparation and a recombinant Saccharomyces cerevisiae strain. World Journal of Microbiology and Biotechnology, 33(7), 1–8.
6. Castillo, F. J., (1995). Organism selection. Bioreactor System Design, 13–45.
7. Chandler, J. P., (1987). Factors influencing monoclonal antibody production in mouse ascites fluid. Commercial Production of Monoclonal Antibodies: A Guide for Scale-Up (pp. 75–92). Marcel Dekker, Inc., New York.
8. Chuong, C. M., Nickoloff, B. J., Elias, P. M., Goldsmith, L. A., Macher, E., Maderson, P. A., & Christophers, E., (2002). What is the 'true' function of skin?. Experimental Dermatology, 11(2), 159–187.
9. Collett, H., (1851). On the treatment of habitual or chronic constipation. Provincial Medical and Surgical Journal, 15(23), 625.
10. Contreras, A., Hidalgo, C., Henschke, P. A., Chambers, P. J., Curtin, C., & Varela, C., (2014). Evaluation of non-Saccharomyces yeasts for the reduction of alcohol content in wine. Applied and Environmental Microbiology, 80(5), 1670–1678.
11. Folbergrova, J., Ljunggren, B., Norberg, K., & Siesjo, B. K., (1974). Influence of complete ischemia on glycolytic metabolites, citric acid cycle intermediates, and associated amino acids in the rat cerebral cortex. Brain Research, 80(2), 265–279.

12. Fröhlich, E., (2018). Comparison of conventional and advanced in vitro models in the toxicity testing of nanoparticles. Artificial Cells, Nanomedicine, and Biotechnology, 46, 1091–1107.
13. Fujita, Y., & Hara, Y., (1985). The effective production of shikonin by cultures with an increased cell population. Agricultural and Biological Chemistry, 49(7), 2071–2075.
14. Gklinos, P., Papadopoulou, M., Stanulovic, V., Mitsikostas, D. D., & Papadopoulos, D., (2021). Monoclonal antibodies as neurological therapeutics. Pharmaceuticals, 14(2), 92.
15. Gresham, R., & Inamine, E., (1986). Nutritional improvement of process. Manual of Industrial Microbiology and Biotechnology (pp. 41–48). Washington: ASM.
16. Gualdrón-López, M., Brennand, A., Hannaert, V., Quiñones, W., Cáceres, A. J., Bringaud, F., & Michels, P. A., (2012). When, how and why glycolysis became compartmentalized in the Kinetoplastea: A new look at an ancient organelle. International Journal for Parasitology, 42(1), 1–20.
17. Holbrook, M. B., (2003). Adventures in complexity: An essay on dynamic open complex adaptive systems, butterfly effects, self-organizing order, coevolution, the ecological perspective, fitness landscapes, market spaces, emergent beauty at the edge of chaos, and all that jazz. Academy of Marketing Science Review, 6(1), 1–184.
18. Jayme, D. W., & Blackman, K. E., (1985). Culture media for propagation of mammalian cells, viruses, and other biologicals. Advances in Biotechnological Processes, 5, 1–30.
19. Junker, B. H., & Wang, H. Y., (2006). Bioprocess monitoring and computer control: Key roots of the current PAT initiative. Biotechnology and Bioengineering, 95(2), 226–261.
20. Keim, C. R., (1983). Technology and economics of fermentation alcohol: An update. Enzyme and Microbial Technology, 5(2), 103–114.
21. Kubicek, C. P., Röhr, M., & Rehm, H. J., (1985). Citric acid fermentation. Critical Reviews in Biotechnology, 3(4), 331–373.
22. Kunkee, R. E., (1984). Selection and modification of yeasts and lactic acid bacteria for wine fermentation. Food Microbiology, 1(4), 315–332.

23. Lee, M., & Phillips, R. S., (1992). Enzymatic synthesis of chloro-L-tryptophans. Bioorganic and Medicinal Chemistry Letters, 2(12), 1563–1564.

24. Lei, L., Yang, C., Patton, M. J., Smelkinson, M., Dorward, D., Ma, L., & Caldwell, H. D., (2021). A chlamydial plasmid-dependent secretion system for the delivery of virulence factors to the host cytosol. Mbio, 12(3), e01179–21.

25. Li, M., Chen, Z., Zhang, X., Song, Y., Wen, Y., & Li, J., (2010). Enhancement of avermectin and ivermectin production by overexpression of the maltose ATP-binding cassette transporter in Streptomyces avermitilis. Bioresource Technology, 101(23), 9228–9235.

26. Maiorca, R., Brunori, G., Zubani, R., Cancarini, G. C., Manili, L., Camerini, C., & Gelatti, U., (1995). Predictive value of dialysis adequacy and nutritional indices for mortality and morbidity in CAPD and HD patients: A longitudinal study. Nephrology Dialysis Transplantation, 10(12), 2295–2305.

27. Marino, G., & Gaggìa, F., (2013). Melia azedarach fruit extracts for control of bacterial contaminants development in micropropagation. Pharmacologyonline, Archives, 3, 128–131.

28. Martin, C. E., & Scheinbach, S., (1989). Expression of proteins encoded by foreign genes in Saccharomyces cerevisiae. Biotechnology Advances, 7(2), 155–185.

29. Martin, Y., & Vermette, P., (2005). Bioreactors for tissue mass culture: Design, characterization, and recent advances. Biomaterials, 26(35), 7481–7503.

30. Matharu, Z., Bandodkar, A. J., Gupta, V., & Malhotra, B. D., (2012). Fundamentals and application of ordered molecular assemblies to affinity biosensing. Chemical Society Reviews, 41(3), 1363–1402.

31. Milsom, P. E., (1987). Organic acids by fermentation, especially citric acid. In: Food Biotechnology (Vol. 4, No. 1, pp. 273–307).

32. Mir, M., Ali, M. N., Barakullah, A., Gulzar, A., Arshad, M., Fatima, S., & Asad, M., (2018). Synthetic polymeric biomaterials for wound healing: A review. Progress in Biomaterials, 7(1), 1–21.

33. Mirón, A. S., Gomez, A. C., Camacho, F. G., Grima, E. M., & Chisti, Y., (1999). Comparative evaluation of compact photobioreactors for

large-scale monoculture of microalgae. In: Progress in Industrial Microbiology (Vol. 35, pp. 249–270).

34. Niaz, B., Nadeem, S., Muzammil, H. M., Khan, J. A., & Zahoor, T., (2009). Optimization of fermentation conditions for enhanced glutamic acid production by a strain of Corynebacterium glutamicum NIAB BNS-14. Pakistan Journal of Zoology, 41(4).
35. Oh, D. J., Choi, S. K., & Chang, H. N., (1994). High-density continuous cultures of hybridoma cells in a depth filter perfusion system. Biotechnology and Bioengineering, 44(8), 895–901.
36. Posillico, E. G., (1986). Microencapsulation technology for large-scale antibody production. Bio/Technology, 4(2), 114–117.
37. Prokop, A., Prokop, Z., Schaffer, D., Kozlov, E., Wikswo, J., Cliffel, D., & Baudenbacher, F., (2004). NanoLiterBioReactor: Long-term mammalian cell culture at nanofabricated scale. Biomedical Microdevices, 6(4), 325–339.
38. Queener, S., & Swartz, R., (1979). Penicillins: Biosynthetic and semisynthetic. Economic Microbiology, 3, 35–122.
39. Rajarathnam, S., Bano, Z., & Steinkraus, K. H., (1989). Pleurotus mushrooms: Part III. Biotransformations of natural lignocellulosic wastes: Commercial applications and implications. Critical Reviews in Food Science and Nutrition, 28(1), 31–113.
40. Remaut, E., Stanssens, P., & Fiers, W., (1981). Plasmid vectors for high-efficiency expression controlled by the PL promoter of coliphage lambda. Gene, 15(1), 81–93.
41. Romano, P., Suzzi, G., Comi, G., & Zironi, R., (1992). Higher alcohol and acetic acid production by apiculate wine yeasts. Journal of Applied Bacteriology, 73(2), 126–130.
42. Ryu, D. D., & Hospodka, J., (1980). Quantitative physiology of Penicillium chrysogenum in penicillin fermentation. Biotechnology and Bioengineering, 22(2), 289–298.
43. Sanchez, S., & Demain, A. L., (2008). Metabolic regulation and overproduction of primary metabolites. Microbial Biotechnology, 1(4), 283–319.
44. Sanders, D., (1981). Physiological control of chloride transport in Chara corallina: II. The role of chloride as a vacuolar osmoticum. Plant Physiology, 68(2), 401–406.

45. Schmidt, C., (1989). The purification of large amounts of monoclonal antibodies. Journal of Biotechnology, 11(2–3), 235–251.
46. Seaver, S. S., (1987). Culture method affects antibody secretion of hybridoma cells. Commercial Production of Monoclonal Antibodies, 4, 49–71.
47. Shamil, S., Birch, G. G., Mathlouthi, M., & Clifford, M. N., (1987). Apparent molar volumes and tastes of molecules with more than one sapophore. Chemical Senses, 12(2), 397–409.
48. Solomons, G. L., (1983). Single-cell protein. In: Critical Review in Biotechnology (Vol. 1, p. 21).
49. Taherzadeh, M. J., & Karimi, K., (2007). Acid-based hydrolysis processes for ethanol from lignocellulosic materials: A review. BioResources, 2(3), 472–499.
50. Taylor, C. R., Shi, S. R., Barr, N. J., & Wu, N., (2013). Techniques of immunohistochemistry: Principles, pitfalls, and standardization. Diagnostic Immunohistochemistry, 2, 1–42.
51. Tejayadi, S., & Cheryan, M., (1995). Lactic acid from cheese whey permeate. Productivity and economics of a continuous membrane bioreactor. Applied Microbiology and Biotechnology, 43(2), 242–248.
52. Tiwari, K., & Gupta, R. K., (2012). Rare actinomycetes: A potential storehouse for novel antibiotics. Critical Reviews in Biotechnology, 32(2), 108–132.
53. Vallino, J. J., & Stephanopoulos, G., (2000). Metabolic flux distributions in Corynebacterium glutamicum during growth and lysine overproduction. Biotechnology and Bioengineering, 67(6), 872–885.
54. Velez, D., Miller, L., & Macmillan, J. D., (1989). Use of tangential flow filtration in perfusion propagation of hybridoma cells for production of monoclonal antibodies. Biotechnology and Bioengineering, 33(7), 938–940.
55. Véliz, R. V., (2002). Alternative techniques to obtain monoclonal antibodies at a small scale: Current state and future goals. Biotecnología Aplicada, 19, 119–131.
56. Vick, B., & Beevers, H., (1978). Fatty acid synthesis in endosperm of young castor bean seedlings. Plant Physiology, 62(2), 173–178.
57. Von Wedel, R. J., (1987). Mass culture of mouse and human hybridoma cells in hollow-fiber culture. Commercial Production of Monoclonal Antibodies, 5, 159–173.

58. Watve, M. G., Tickoo, R., Jog, M. M., & Bhole, B. D., (2001). How many antibiotics are produced by the genus Streptomyces?. Archives of Microbiology, 176(5), 386–390.
59. Werner, R. G., Kopp, K., & Schlueter, M., (2007). Glycosylation of therapeutic proteins in different production systems. Acta Paediatrica, 96, 17–22.
60. Wurm, F. M., (2004). Production of recombinant protein therapeutics in cultivated mammalian cells. Nature Biotechnology, 22(11), 1393–1398.
61. Xu, D., Tan, Y., Li, Y., & Wang, X., (2011). Construction of a novel promoter-probe vector and its application for screening strong promoter for Brevibacterium flavum metabolic engineering. World Journal of Microbiology and Biotechnology, 27(4), 961–968.
62. Zhang, X., Tu, B., Dai, L. R., Lawson, P. A., Zheng, Z. Z., Liu, L. Y., & Cheng, L., (2018). Petroclostridium xylanilyticum gen. nov., sp. nov., a xylan-degrading bacterium isolated from an oilfield, and reclassification of clostridial cluster III members into four novel genera in a new Hungateiclostridiaceae fam. nov. International Journal of Systematic and Evolutionary Microbiology, 68(10), 3197–3211.

Chapter 2

Microorganisms in the Fermentation Process

CONTENTS

2.1. INTRODUCTION

The word "fermentation" is derived from the Latin term "fever," its meaning is "to boil." Fermentation had been described by Louis Pasteur as "La vie sans l'air," or " without air, life is possible" and the fermentation science is also referred to as zymurgy or zymology. On the other hand, fermentation is among the oldest known food preservation or storage technologies all over the world going back to the Period of the Neolithic (10,000 years before Christ) (Joshi et al., 2021; Voidarou et al., 2021). In a sense of biochemistry, it is the metabolic process that breaks down organic complex molecules (particularly sugars) into smaller ones without the use of oxygen (exogenous oxidizing agent), and energy is also generated (ATP molecules).

Fermentation's byproducts vary based on the type of bacteria used. The scientific stimulus for fermentation began with the identification and naming of microbes by Leeuwenhoek and Hooke in 1665. Following that, in 1859, Pasteur destroyed the "spontaneous generation theory" by well-planned experiments. On the other hand, Sir Joseph Lister investigated the effectiveness of a unique bacterium Lactococcus lactis in milk that had been fermented in 1877 (Santer, 2010). In most cases, microbes that take the shape of a cluster of cells or a single cell, such as algae, bacteria, fungus, or plant or animal cells, are engaged in the fermentation process. There are many factors in the process of fermentation, for example, the proportion of microorganisms and their components (enzymes), as well as parameters like temperature, pH, and fermented medium (anaerobic or aerobic). Generally, microorganism's fermentation is classified into 4 categories: (i) metabolite production; (ii) biomass generation (feasible cellular material); (iii) the substrate conversion or transformation into value-added products; and (iv) vitamin, enzyme, and protein synthesis; (Thakur et al., 2015). The fermentation by microorganisms, their development and metabolism, factors that have an impact on fermentation and bio-products (metabolites, both primary and secondary) obtained from the process of fermentation, and the use of probiotics in the fermentation of food are all covered in this chapter.

2.2. DIVERSITY OF MICROBES

There is no broad consensus on the subject of classification/categorization of microbes, and categorization is sometimes random. Eukaryotic and Prokaryotic cells are the two basic types of cells. The absence or existence of a membrane the genetic substance of the cell is the major distinction or characteristic among them (Amna et al., 2006).

With several chromosomes (DNA molecules) in the nucleus, the inner structure of eukaryotes is more complicated. Eukaryotic cells are made of a genuine nuclear membrane that contains Organelles bounded with the membrane (such as the mitochondria, endoplasmic reticulum (ER), and the Golgi apparatus) as well as a range of specialized and functional organelles.

A single chromosome and a basic structure is present in Prokaryotes and they are unicellular organisms. Nuclear membrane and membrane-bound organelles (ER and mitochondria) are absent in prokaryotic cells. Table 2.1 provides a comprehensive comparison between eukaryotes and prokaryotes.

Table 2.1. Contrast Among Eukaryotes and Prokaryotes (Gomez et al., 2015)

Characteristics		**Eukaryotes**	**Prokaryotes**
Genome	DNA molecules quantity	Several	One
	Organization of genetics (DNA)	Chromosome located in the nucleus of a membrane	Chromosome abundantly located in a cytoplasmatic area termed the nucleoid
		DNA is straight and chromatin-forming proteins	DNA is circular, without protein
		Bound with histone proteins	Bared such as not protein-bound and thus does not make chromatin
	Plasmid	But organelles might have their chromosomes, such as mitochondria	Contain extra-chromosomal plasmids
	Organelle DNA	Yes (mitochondrial and chloroplastic gene)	No
	Cell division	Mitosis or meiosis	Binary fission/budding
	Nuclear membrane	Present	Not present

Organelles	Endoplasmic reticulum	Present	Not present
	Mitochondria	Present	Not present
	Golgi apparatus	Yes	No
	Ribosome	80 S	70 S
	Photosynthetic apparatus	Chloroplasts (plants)	Chlorosomes
	Flagella	Complicated structure, with microtubules	Simple structure, single protein
Spores	Types	Both exospores and endospores	Endospores
	Resistance to heat	Low	Elevated
Example	–	Fungi, protozoa, algae, and helminths	Bacteria and archaea

According to facts, there are three inherent or distinctive branches of life (eubacteria (or "real bacteria"), archaebacteria, and eukaryotes) that have been created through a common/universal predecessor (family history) (Table 2.2).

Table 2.2. Fulfills with Certain Features Inherent in Such Groups

Group	Constituent Groups	Properties	Cell Structure
Eubacteria	Most of the bacteria	Similar to eukaryotes	Prokaryotic
Eukaryotes	Plants (such as ferns, the seed of plants, mosses)	Multicellular	Eukaryotic
Archaebacteria	Thermophiles, methanogens, halophiles	Distinctive properties	Prokaryotic

2.2.1. Prokaryotes

The Optimum or the most prokaryotes have lengths ranging from 0.5 to 3 micrometers (m) (or equivalent radius). The spirillum/spiral (for example, Rhodospirillum) or bacillus/cylinder (for example, Escherichia coli) or coccus/spherical(for example, staphylococci) are examples of different species that have different forms. Prokaryotic cells multiply quickly, with doubling durations ranging from ½ hour to many hours. Furthermore, prokaryotes might obtain energy from a range of carbon sources, including carbohydrates, hydrocarbons, carbon dioxide (CO_2), and proteins (Gomez et al., 2015).

2.2.1.1. Eubacteria

Eubacteria are minuscule single-cell organisms, distinctive of a stiff cell wall, that grow in a variable environment. There are various groups of eubacteria. Eubacteria are a kind of gram-positives in the family naming Eubacteriaceae. The term "gram staining" was derived by the Danish bacteriologist Hans Christian Gram who established the technology in 1884 (Anastassiadis et al., 2002). First of all, the procedure of staining begins with the staining/fixing of the cells (by heating), followed by the addition of Base dye (crystal violet), which results in staining of all of the bacteria to purple. Then iodine (binds crystal violet) is introduced and then ethanol (or acetone) quick addition or decolorization. Gram-positive cells would remain purple, whereas gram-negative cells become colorless. Ultimately, fuchsine or safranin leaves gram-negative cells red, whereas gram-positive cells are purple. The physical and chemical features of the cell wall of the bacteria (or Cell envelope structure) differ bacteria by their capacity to respond with the gram stain (Gomez et al., 2015).

E. coli is a standard gram-negative cell (Figure 2.1). A thin peptidoglycan (commonly referred to as murein) layer supports the outer membrane. Peptidoglycan is a polymer of amino acids and sugars (polysaccharides) with a mesh/chain-link wall structure. Peptidoglycan offers the strength to the structure while also resisting (opposing) the cytoplasm's osmotic pressure. The periplasmic space separates a 2nd membrane (the cytoplasmic or internal membrane) from the external membrane (Asahi et al., 1996). The bacterial plasma membrane (cytoplasmic membrane) is made of bi-layered phospholipids (containing 30% lipids, 20% carbohydrates, and 50% protein).

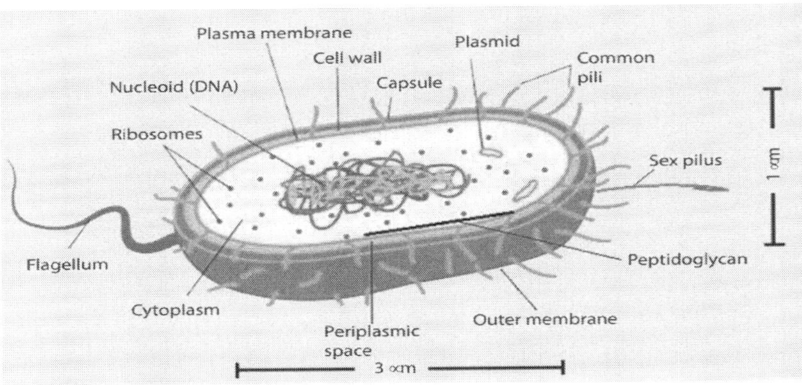

Figure 2.1. A typical cell of gram-negative bacteria.

Source: https://www.chegg.com/flashcards/microbiology-6ddf8a86-88b3-4568-9f94-ba3890a7b092/deck.

The cell membrane (polysaccharide capsule) retains and protects major cellular compounds, in turn; prevent unwanted substances within the cell environment. It supports the integrity of the membrane or structure and when loses the integrity of the membrane leads to cell lysis and the death of the cell. The transfer of desired material in and out of the cell is essential on the cell membrane (Kaul et al., 2000).

Bacillus subtilis is an example of a gram-positive cell. The outer membrane in Gram-positive cells is absent. Rather, they have a thick, stiff cell wall that is composed of the strands of glycan which are cross-connected to side chains of peptide and have many layers of peptidoglycan. Teichoic acids are connected covalently to the peptidoglycan (the acids of wall teichoic) or are connected to the membrane of lipid in Gram-positive cells (lipoteichoic acids). Gram-positive bacteria are generally more suitable for protein extraction since they have a single cytoplasmic membrane. When the protein is a desirable product, such extraction might be technically beneficial. Asymmetric cell division yields physically distinct daughter cells during the forming of spore by gram-positive bacteria (B. subtilis).

Spores have a different cell wall than mother cells and have their collection of protein (Navarre and Crump, 2004). While the DNA of eukaryotic, is packed in the nucleus carefully, the DNA of bacteria flows easily in the cytoplasm as a tangled mass called nucleoid.

Bacteria like actinomycetes (usually gram-positive and anaerobic), although have lengthy filaments and heavily branching hyphae that physically resemble molds. Although, the cell wall components and the lack of a nuclear membrane require categorization like bacteria. Bifidobacterium, Thermomonospora, Microbacterium, Corynebacterium, Streptomyces, Actinomyces, Brachybacterium, Brevibacterium, and Micrococcus are the example of the appropriate generation for this category (Xiao et al., 2010). Antibiotics are obtained from actinomycetes. Streptomyces bacteria have been identified as the producers of over 500 different bioactive metabolites, including antivirals, antibacterials, enzyme inhibitors, and antifungals. Pectinolytic, cellulolytic, and amylolytic enzymes are found in actinomycetes and are helpful in the enzymatic hydrolysis of pectin, cellulose, and starch, accordingly (Kar and Ray, 2008, 2011).

Other Eubacteria features might be derived from energy metabolism and cellular nutrition. Photosynthesis is an excellent example. Cyanobacteria (such as blue-green algae) have chlorophyll and acquire their energy from photosynthesis. They also can set CO_2 into carbohydrates and create oxygen. The fixing of nitrogen by cyanobacteria (set the nitrogen of atmospheric under the conditions of anaerobic) uses specialized cells called heterocysts.

When prokaryotes are observed under a microscope, they may contain various visible structures, like storage granules, ribosomes, volutins, and spores. Ribosomes are sites where biological proteins are produced. Prokaryotes are composed of the 30S and 50S subunits and also contain 70S ribosomes and have highly complicated structures of cellular machinery (Kar and Ray, 2008).

Storage granules (absent from each bacteria) are employed as a container for the maintenance of nutrition in the cytoplasm. Polyphosphates, polysaccharides (glycogen), nitrogen, sulfur granules and lipids are the source of essential metabolites of Storage granules. Volutins are polyphosphate bodies which are composed of inorganic polymetaphosphates, and these are granular intracellular energy-rich compound, which are present in some species (such as Rhodospirillum rubrum and Agrobacterium tumefaciens) (Seufferheld et al., 2011). Protein, lipids, and nucleic acid are sometimes linked with these granules, and they operate as a reserve bank of energy and phosphate. Chromatophores are huge inclusion bodies (light-reflecting and pigment-containing cells) used in photosynthesis for light absorption by certain photosynthetic bacteria (such as Rhodospirillum rubrum) (Kaul et al., 2000).

Intracellular spores, also known as endospores, are also produced by certain bacteria. The endospores of bacteria (structures are like the seed) are hard, inactive, and non-reproductive structures generated by bacteria as a means of resistance to extreme conditions (such as radiation, toxic chemicals, and high temperature). Spores may germinate in the presence of suitable growth conditions, resulting in bacteria that are actively developing. Some bacteria (such as Bacillus antacids, Streptococcus) have a capsule (outside cell wall), which are lipopolysaccharides (polysaccharides) or polypeptide coating (lipoproteins). Several bacterial cells generate a slime-shaped extracellular substance that shields them from changes in the environment (such as desiccation and antibiotics). In contrast, the bacterial capsule is firmly connected and has clear boundaries to the bacterium, lose connection to the slime layer and undergo quick cleanup (Gomez et al., 2015).

2.2.1.2. Archaebacteria

Archaebacteria are related to eukaryotes. However, their molecular level is vastly different from these cells. Both Archaebacteria and Eubacteria, or "genuine" bacteria, are classified as "single-celled" microorganisms and belong to the same kingdom. But advanced molecular and genetic research allowed creating the primary biochemical differences among Eubacteria and Archaebacteria. The following are the distinctions among Eubacteria and Archaebacteria (Table 2.3) (Boyle, 2017):

- Eubacteria live in and on big organisms and are often seen in water and soil, however, Archaebacteria live in harsh conditions (such as ocean depths, hot springs, and salt brines) and also have a unique metabolism.
- Old bacteria are referred to as Archaebacteria, whereas "true" bacteria referred to as Eubacteria.

Table 2.3. Characterization of Different Components of Bacteria (Gomez et al., 2015)

Component	Diameter or Size	Characteristics or Composition
Extracellular Substances		
Capsule	0.5 to 2.0 µm	Sometimes polypeptides (such as Bacillus antracis), mainly polysaccharides (such as Streptococcus).
Slime layer	–	Sometimes polypeptides (such as Bacillus subtilis), mostly polysaccharides (such as Leuconostoc).
Microcapsule	5 to 10 nm	The protein-polysaccharide-lipid combination ensures that enteric and other species are exact antigens.
Cell Membrane		
–	5 to 10 nm	The semipermeable barrier to nutrients; consisting of 50% proteins, 28% lipid and 15–20% carbohydrate in the double-layered membrane.
Cell Wall		
Gram-negative species	10 to 20 nm	It usually consists of a combination of protein-polysaccharides with a lesser quantity of muramic polymer.
Gram-positive species	10 to 20 nm	Rigidity and cell form; mostly comprises polysaccharide mixed polymers and N-acetyl muramic polymers, peptides, and the acids of teichoic.
Pilus		
Fimbria	5 to 10 nm by 0.5 to 2.0 µm	Rigid cell protein testing. E. coli is produced by particularly long ones.
Flagellum		
–	10 to 20 nm by 4 to 12 µm	The lengthy and spiral filaments can be observed on any end. The motility protein of myosin keratin fibrinogen comes from the membrane of the cell.
Nuclear Material		
–	–	Genetically operating DNA-like genes all arranged into one endless chromosome.
Inclusions		
Granule of Storage	0.5 to 2.0 µm	Glycogen like lipid granules, or sulfur can be observed in a few species.
Spore	1.0 to 1.5 µm by 1.6 to 2.0 µm	Intracellularly, each cell produces one spore. Spores demonstrate excellent heat resistance, antibacterial agents, and dryness resistance.

Volutin (polyphosphate bodies)	0.5 to 1.0 µm	Metaphosphates are inorganic
Ribosome (70S)	10 to 30 nm (approx. 20 nm or 200 A°)	Organelles are essential for protein production; approximately 10.000 ribosomes in each cell; consist of 37% ribosomal protein and 63% rRNA.
Chromatophore	50 to 100 nm	Photosynthesizing species organelles (such as Rhodospirillum rubrum).

Note: S: Svedberg rate or unit of sedimentation; rRNA: ribosomal ribonucleic acid; DNA: deoxyribonucleic acid.

- The sequences of nucleotide in the rRNA are identical with Archaebacteria and distinct from those of Eubacteria.
- Peptidoglycan is not observed in Archaebacteria.
- Eubacteria are classified as gram-positive and gram-negative, however, Archaebacteria are classified as methanogens (methane-producing bacteria), halophiles (bacteria that thrive in highly salty solutions), and thermophiles (grow at decreased pH and high temperatures).
- The cytoplasmic membrane lipid content differs significantly between the two groups (Gomez et al., 2015).

2.2.1.3. Proteobacteria

The phylum Proteobacteria is one of the most important phyla of gram-negative bacteria. Gluconacetobacter sp. and Acetobacter sp. are the two major bacteria groupings in this phylum, with 8 and 9 species, respectively. Most of them are used in the production of vinegar, although they are also important in cocoa and coffee fermentation (Sengun and Karabiyikli, 2011).

2.2.1.4. Eukaryotes

Eukaryotes are made up of protozoa, algae, fungi (molds and yeasts), and plant and animal cells. Eukaryotes are 5 to 10 times in diameter to that of prokaryotes (e.g., yeast has a diameter of approximately five µm, animal cells have a diameter of about 10 µm, and plant cells have a diameter of about 20 µm). Inside the cytoplasm of eukaryotes have many cellular organelles (for example, Golgi apparatus, mitochondria, etc.), and a genuine nucleus (contained within the membrane) (Gomez et al., 2015). Two typical eukaryotic cells are depicted schematically in Figure 2.2.

Eukaryotes and prokaryotes have distinct cell walls and cell membrane structures. Phospholipids and Proteins create a bilayer structure in the plasma membrane (lipid bilayer with entrenched protein). The membrane's major proteins are hydrophobic and are encased in a matrix of phospholipids (which make about half of all lipids). All cytoplasmic membranes or eukaryotic cells require sterols (Nicolson and Ash, 2017). The presence of sterols (3rd main type of lipid membrane after glycolipids and phospholipids) in the plasma (plasmalemma) membrane of eukaryotes has a significant distinction.

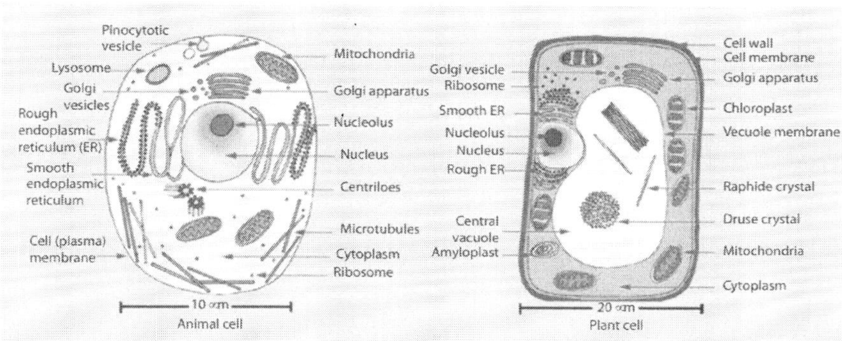

Figure 2.2. Two eukaryotic cells schematic description.

Source: https://courses.lumenlearning.com/suny-biology1/chapter/eukaryotic-cells/.

Cholesterol, for example, is the most abundant sterol in mammalian plasma membranes. Sterols improve membrane stiffness and strength (Low fluidity and permeability of the membrane) while making the covering less elastic (Gomez et al., 2015).

The eukaryotic cell walls (for example, plants, algae, and fungus) differ significantly. Cellulose and polysaccharides are found in the cell walls of certain eukaryotes (such as algae), whereas chitin and other polysaccharides (such as algae) are found in the cell walls of other eukaryotes. The cell walls of true fungus do not include cellulose. The cell walls of the plants consist of cellulose fibers linked via adherence to hemicellulose to create a system that is incorporated into pectin aggregates. The cell wall is strengthened by the cellulose-hemicellulose-pectin network. A cell wall does not exist in animal cells; instead, they have a plasma/cytoplasmic/cell membrane. According to a theory, animal cells are fragile (delicate) and extremely shear-sensitive (Gomez et al., 2015). This complicates the structure of wide-scale bioreactors for the animal cells.

The eukaryotic nucleus (excluding animal red blood cells) is surrounded by a membrane that carries chromosomes like nuclear material (DNA-linked proteins). The double membrane that makes up the nuclear membrane/envelope is comprised of porous (nuclear holes) and concentric membranes. The nucleus is fully encased in the nuclear envelope, which isolates the genetic substance of the cell from the cytoplasm. Pores of the nucleus are necessary for controlling the nuclear movement of molecules around the wrapping (Mazzanti et al., 2001). The nucleolus is a stained region inside the nucleus that is where rRNA and ribosome synthesis takes place. It is not encased in a membrane. Despite this, several chromosomes contain a lesser quantity of ribonucleic acid and fundamental proteins termed as histones attached to the deoxyribonucleic acid (DNA). The histones are linked to a single linear DNA molecule on each chromosome (Gomez et al., 2015).

Mitochondria are the strongholds for a eukaryotic cell, where oxidative phosphorylation and respiration take place. They are double membrane-bound organelles that create the majority of the cell's supply of adenosine triphosphate (ATP). Mitochondria have a roughly cylindrical form with a diameter of 1 μm and a length of 2–3 μm (size and structure vary greatly). Figure 2.3 depicts the regular organization of mitochondria. Two layers of phospholipid-containing protein incorporated in the matrix of lipid make up the exterior membrane. The ratio of protein-to-phospholipids is similar to the plasma membranes of eukaryotes (one-by-one ratio, approximately). Cristae are a complex system of partitioned inner membranes found in mitochondria. The area inside the cristae is filled with a gel-like matrix holding vast quantities of protein (enzyme combinations).

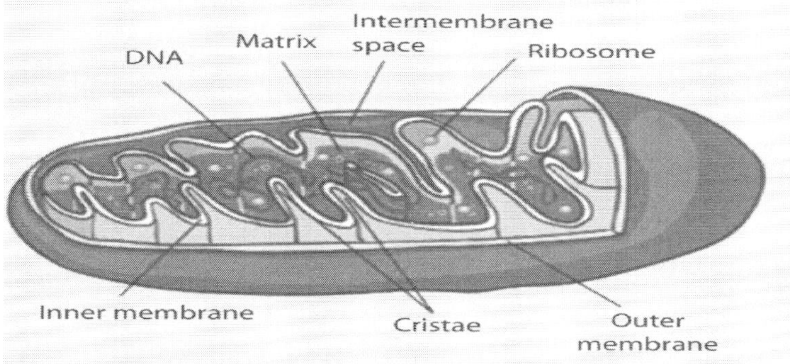

Figure 2.3. The regular structure of mitochondria.

Source: https://www.medicalnewstoday.com/articles/320875.

Certain oxidative respiration enzymes (involved in fatty acid oxidation, pyruvate oxidation, and the cycle of citric acid) are linked to the cristae. Mitochondria contain DNA that is structured as many copies of a cell, typically circular chromosome, as well as protein-synthesizing machinery and the ability to reproduce autonomously (Gomez et al., 2015).

The ER is a system of the tangled and complicated membrane that is made up of an interconnected network of tubelike/flattened structures called cisternae. Smooth endoplasmic reticulum (SER) and rough endoplasmic reticulum (RER) are the two kinds of ER. The RER (granular) includes ribosomes on the external surface (it gives an irregular look) and have the location for the making of protein and post-synthesis modifications of the design of the protein. The SER (agranular) is more concerned with steroids production, phospholipids, and lipids (Dallner et al., 1966).

Lysosomes are membrane-bound organelles that store and release digestive and hydrolytic enzymes. The digestion of nutrients and invading substances or biomolecules (for example, peptides, carbohydrates, lipids, and nucleic acids) is aided by lysosomes.

Peroxisomes are structurally identical to lysosomes but not functionally. There are oxidative enzymes (enzymes that carry out oxidative processes) in peroxisomes that generate H_2O_2.

Glyoxysomes (precise peroxisomes of the plant) are quite tiny membrane-bound particles that carry the cycle of glyoxylate in the enzymes.

The bodies of Golgi (the apparatus of Golgi) are cytoplasmic endomembrane systems made up of membrane aggregates or combinations that are involved in the ejection of some protein. The bodies of Golgi are places where protein (glycosylation of enzymes) is changed through adding sugar monomers in a process known as glycosylation. Protein function (responsibility) in the organism requires such modification/conversion.

Vacuoles are the bounded membrane of organelles (full by the sap of the cell) with a less density (no fundamental form) that are surrounded through the tonoplast membrane. Osmotic control (hydrostatic pressure inside the cell), the storage of waste products, and food digestion are all handled by vacuoles. Vacuoles can take up a substantial portion of a cell's volume (approximately 90% in the cells of the plant) and keep the interior pH acidic (Tang et al., 2020).

In photosynthetic eukaryotes (plant and algae cells), chloroplasts (also called plastid) are fairly big organelles having chlorophyll that are essential for the process of photosynthesis. Each chloroplast has two membranes, an exterior membrane and a vast amount of internal membranes known as thylakoids. Thylakoid contains a normal structure of the membrane with two layers of lipid, coupled with chlorophyll molecules. Chloroplasts are self-contained units with their DNA (ctDNA) and protein-synthesis machinery.

Flagella are lengthy, filaments-like structures that are linked to either side of the cell and are important for cell movement in some prokaryotes and eukaryotes. Two core fibers are enclosed by 18 (9 merged pairs of microtubules) peripheral fibers that occur in pairs in eukaryotic flagella (Khan and Scholey, 2018). Fibers are made up of proteins called tubulin and are arranged in a structure of tubes called microtubules. A membrane surrounds the entire fiber assembly, which is assembled inside an organic matrix. Cilia are flagella-like structures that are smaller and more numerous than flagella. Cilia are found only in one kind of protozoa, also known as the ciliates (such as Paramecium). Eukaryotic cilia are physically similar to eukaryotic flagella; however, they are distinguished by length or function. Flagellated organisms are significantly slower than ciliated organisms in terms of movement. In eukaryotic cells, the cytoskeleton relates to filaments that serve as an internal framework for controlling its form and organizing the cell's internal processes. Such filaments are important for the motion of cell and chromosomal segregation into 2 daughter cells during the division of the cell, as well as the conversion of mechanical pressures into biological responses. Intermediate filaments, microtubules, and actin filaments are the three kinds of fibers found (Fife et al., 2014).

2.2.1.5. Microscopic Eukaryotes

Fungi are heterotrophs (organisms that acquire their nourishment by the absorption of dissolved molecules) that have a broad variety of habitation and distribution. Fungus cells (such as fungus mycelia) are bigger than the cells of the bacteria or its inner structures, such as vacuoles and nuclei. The fungal mycelia (fungus) may be observed with the human eye and with a light microscope. Yeasts and molds are two major categories of fungi.

Yeasts are eukaryotic, unicellular microorganisms with a diameter of 5 to 10 micrometers. Yeast cells may be spherical, oval, or cylindrical. Yeasts may reproduce in two ways: sexually and asexually. Fission and budding are the two methods of asexual reproduction (growth of vegetative). A tiny bud (also known as a "daughter cell" or "bleb") develops on the parent cell during budding, which slowly enhances and splits from the parent cell. Fission-based asexual reproduction is as in the case of bacteria. Some species of yeast (such as Saccharomyces pombe) are capable of fission reproduction (Acosta et al., 2011). Fission occurs when a cell reaches a definite size and then divides into two identical cells. The development of a zygote (a diploid cell) from the combination of the two haploid cells, each with one pair of chromosomes, is the basis of sexual reproduction in yeasts. Ascospores are formed when the nucleus of diploid cells splits multiple times. Every ascospore eventually develops into a new haploid cell capable of fission or budding reproduction. Figure 2.4 depicts the life cycle of a normal yeast cell. The nutritional requirement of cells and the reproductive mechanisms (such as fission or budding) are used to classify yeasts (Vyas et al., 2021).

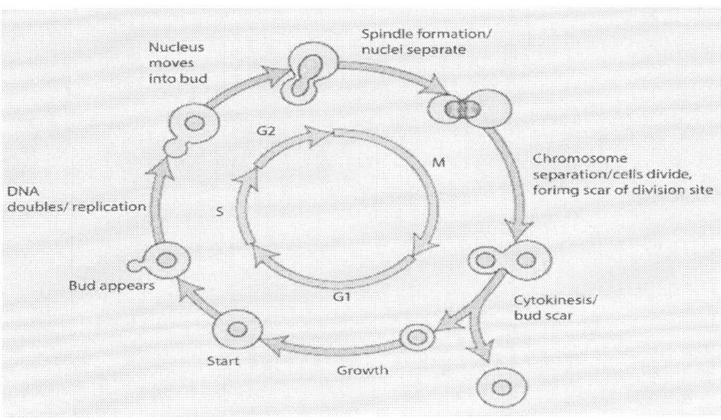

Figure 2.4. The life cycle of normal yeast cell.

Source: https://microbialcellfactories.biomedcentral.com/articles/10.1186/s12934-018-1044-2.

The yeast Saccharomyces cerevisiae is most commonly utilized and is one of the most common microorganisms used in industries including baking, distilling, and winemaking. Under the conditions of less oxygen (for example, for the production of alcoholic drink, whisky, and alcohol), Saccharomyces cerevisiae is utilized in the production of alcohol and also in the formation of bakery under the environment of low oxygen, in which it transforms fermented sugar (found in bread) into CO_2 (Gomez et al., 2015).

Molds are filament fungus with a mycelial structure (such as a tubular branching network). It is a system of extremely branched (mycelium) tubes containing several nuclei and cytoplasm. Hyphae are multicellular filaments on the mycelium that are thin and long. Mycelium may form in the atmosphere, and conidia are recognized as the asexual spores (division at the terminal of the hyphae). Conidia have a roughly round shape and are usually colored. Molds may reproduce sexually and produce sexual spores (Ribes et al., 2000). These spores are resistant to freezing, drying, heat, and several chemical agents. Mold spores may germinate and produce hyphae in both asexual and sexual forms. Mold structure and asexual reproduction are shown in Figure 2.5. Molds grow readily on damp places, solid nutrient surfaces and produce long, highly branched cells. In industrial fermentation operations, the development of molds in the form of pellets may be an option. Mold development in submerged culture frequently results in the aggregates of cells and pellets. After fermentation, the pellets have a diameter of 1 mm to 2 mm. Pellet formation may create issues with nutrition transport (especially oxygen) within the granule. Pelletization, on the other hand, decreases broth viscosity, which may help in bulk oxygen transport. The citric acid is produced using molds (Aspergillus niger), and also pellet growth is commonly applied in the manufacturing of citric acid. Mold (Penicillium notatum-chrysogenum group) produces many antibiotics, including penicillin (Gomez et al., 2015).

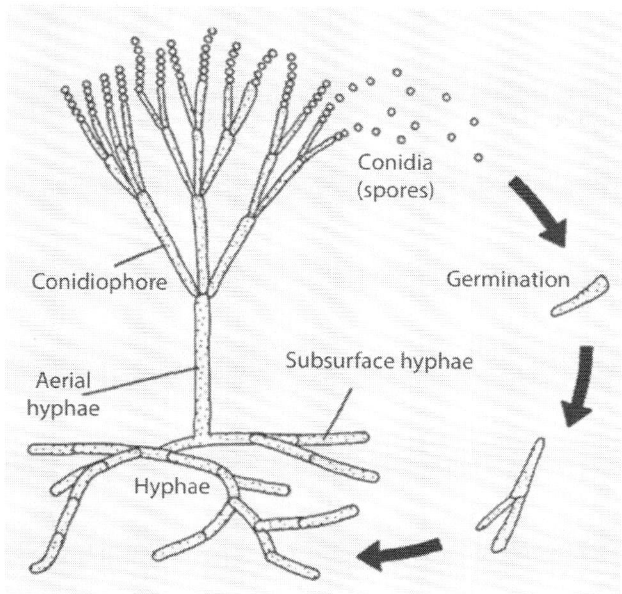

Figure 2.5. Asexual reproduction and structure of molds.

Source: https://www.cliffsnotes.com/study-guides/biology/microbiology/the-fungi/structure-and-physiology-of-fungi.

Algae are generally unicellular (eukaryotic) organisms (for example, diatoms, and chlorella). A normal single-celled alga has a size between 10 to 30 μm. Multicellular algae (such as stonewort, spirogyra, and giant kelp), although, occasionally produce an unbranched or branched filament structure. In marine water, there exist certain algae having a multicellular structure (for example, seaweeds) (Thompson et al., 2012).

Algae are photosynthetic and contain chloroplasts (organisms with a design similar to cyanobacteria), which give them a color of green. The chloroplasts are essential for photosynthesis as they contain chlorophyll pigments. The cell walls of certain algae include silica or calcium carbonate. On the industrial scale, silica in the cell walls of diatoms is utilized as filtration aids. Spirulina, Chlorella, Dunaliella, and Scenedesmus algae are utilized for wastewater treatment (WWT) and single-cell protein synthesis, simultaneously. Seaweeds and marine algae provide some gelling agents (for example, alginate or alginic acid). Because of the presence of additional pigments, certain algae appear red or brown (Kaul et al., 2000; Gomez et al., 2015).

Protozoa are filamentous, single-celled, parasitic, or free-living eukaryotic cells with no cell walls and are rather big in size (1 to 50 mm). Protozoa are heterotrophic, so they eat organic particles or other tiny creatures for nourishment. Protozoa are generally single-celled and reproduce either asexually or sexually (Matz and Kjelleberg, 2005). They are categorized based on how they move. The amoeba moves in an amoeboid motion, which involves the cell's cytoplasm flowing onward to create a pseudopodium, and the remainder of the cell flowing in the direction of this lobe. The flagellates are mobile with their flagella. Flagella helps trypanosomes travel around and induce a variety of illnesses in people. The ciliates move via the motion of cilia, which are tiny arms on the surface of the cell. Food particles are not ingested by such protozoa; instead, dissolved food components are absorbed by their membranes. Animal and human parasites are found among the sporozoans, which are non-motile. Giardiasis, amoebic dysentery, and malaria are all illnesses caused by protozoa (a human parasite). Some protozoa are useful in biological WWT technologies for killing toxins and producing cleaned effluent (Gomez et al., 2015).

2.3. METHODS OF FERMENTATION

Fermentation is divided into two types: LF (liquid fermentation) or SMF (submerged fermentation), and SSF (solid-state fermentation) based on the amount of water utilized throughout the process.

2.3.1. Submerged Fermentation (SMF) or Liquid Fermentation (LF)

SMF takes advantage of cheap liquid substrates, molasses, and broths. In the fermentation broth, bioactive substances are released. Because the substrates are quickly depleted, they must be renewed or supplied with nutrients regularly. This fermentation technique is ideal for microorganisms that require a high moisture content, such as bacteria. Purification and refining of goods are also made easier using this technique/method. SMF is primarily utilized in the removal of secondary metabolites that need liquid extraction (Kaksonen et al., 2018).

2.3.2. Solid-State Fermentation (SSF)

Solid substrates such as paper pulp, bran, and bagasse are used for SSF. The basic benefit of adopting such substrates is that waste substances having excess nutrition may be recovered quickly and efficiently like substrates.

A particular substrate may be utilized for lengthy fermentation durations and may be utilized gently and gradually in this fermentation method. This method will now facilitate the regulated release of nutrients. SSF is most suited for fermentation processes that rely on low moisture content, i.e., microorganisms such as fungus (Behera and Ray, 2016). However, it cannot be utilized in fermentation procedures involving other organisms or bacteria that need a high-water activity (Table 2.4) (Babu and Satyanarayana, 1996).

Table 2.4. Comparison between Solid and Liquid Substrate Fermentation (Mienda et al., 2011)

Factor	Solid Substrate Fermentation	Liquid Substrate Fermentation
Aseptic conditions	Non-sterile conditions, vapor treatment	Aseptic control and heat sterilization
Substrates	Polymer insoluble substrates: Cellulose, starch, lignin pectin,	Soluble substrates (sugars)
Metabolic heating	Heat transfer capacity is low	Easy temperature management
Water	Less water use; low water activity. Effluent is not present	Excess water utilized and effluents released
pH control	Buffered solid substrates	Easy pH control
Aeration	Simple aeration and elevated surface exchange air	Low soluble oxygen. A large amount of air needed
Scale-up	Need for new design equipment and engineering	Commercial equipment existing
Mechanical agitation	Static conditions preferred	Good homogenization
Contamination	Risk of pollution for the low growth rate of fungi	Risks of pollution for one strain of bacteria
Inoculation	Spore immunization, batch	Easy immunization, a nonstop process
Volume of equipment	Fewer volumes and fewer costs of equipment	Elevated volumes and elevated cost of technology
Energetic consideration	Less consumption of energy	Elevated consumption of energy
The concentration of substrate/products	100 or 300 grams per liter	30 to 80 grams per liter
Effluent and pollution	No waste matter, low contamination	High waste matter, large contamination

Yeasts and bacteria both are engaged in processes of SMF and SSF, but fungi are often engaged in SSF activities. The functions of yeasts and bacteria in SMF are frequently linked to the foodstuff and drink processing sectors and would be addressed with further depth in the next sections. Because of their metabolic, physiological, and enzymological characteristics, as well as their abundance in oriental foodstuffs, composting processes, and ensiling, filamentous fungi are most suited for SSF (Table 2.5) (Shaktimay et al., 2010).

Table 2.5. Examples of Key Groups of Microbes Concerned with SSF and SMF Processes

Microorganisms	SMF	SSF
Bacteria		
Clostridium sp.	Pesticide degradation	Ensiling, food
Bacillus sp.	Enzymes (polygalcturonase, α-amylase, phytase, etc.)	Natto, composting, α-amylase
Serratia sp.	–	Composting
Pseudomonas sp.	Xenobiotic degradation	Composting
Lactic acid bacteria	Fermented foodstuff (pickle, yogurt, sauce, etc.)	Ensiling, food
Fungi		
Penicilium notatum, roquefortii	Penicillin	Cheese
Alternaria sp.	–	Composting
Pleurotus ostreatus, Sajor-caju	–	Mushroom
Lentinus edodes	–	Shiitake mushroom
Amylomyces rouxii	–	Cassava tape
Aspergillus niger, A. oryzae	Foodstuff, enzyme (like amylopullulanase)	Foodstuff, enzyme (like glucoamylase)
Phanerochaete chrysosporium	–	Lignin degradation, composting
Beauveria sp., Metharizium sp.	Bio-insecticide	Bio-insecticide
Trichoderma sp.	Cellulase	Composting biological control, bioinsecticide

Rhizopus spp.	Foodstuff, enzymes, natural acid	Composting, foodstuff, enzymes, natural acid
Yeasts		
Saccharomyces cerevisiae	Ethanol, alcoholic drinks	Ethanol, alcoholic drinks
Endomicopsis burtonii, Schwanniomyces Castelli	–	Cassava tape

Fungi are more effective for bioconversion of solid substrates because of their hyphal modes of development and their strong tolerance to high osmotic pressure and low water activity conditions. Tempeh (in Indonesia) and Koji (in Japan) are the two most effective filamentous fungi-based SSF applications. The fungus Aspergillus oryzae is cultivated on wheat bran and soybean to produce Koji, which is the 1st stage in the manufacturing of citric acid and soy sauce. Koji is formed following an unclean condition with many hydrolytic enzymes that are necessary for subsequent stages of the process of fermentation. Likewise, Tempeh, a fermented foodstuff made from the development of Rhizopus oligosporus upon soybeans (Shieh and Beuchat, 1982). The hydrolytic enzymes secreted through R. oligosporus destroy non-nutrients including trypsin inhibitors and hemagglutinins (lectins) in crude soybean during Tempeh fermentation and improving zest, flavor, and nutritional worth. Furthermore, filamentous fungi have a benefit against single-celled bacteria in colonizing solid substrate and using resources due to their hyphal mode of development (Behera et al., 2019). A mixture of apical growth of hyphal and the mycelium provides a solid structure in fungus development. The hydrolytic enzymes are discharged near the hyphal tip, rather than being diluted as in SMF.

2.4. MEASUREMENT OF GROWTH DURING FERMENTATION PROCESS

Microorganisms require moisture and a range of other materials such as elements of mineral, development stimulants, and gases, i.e., oxygen, for growth. Almost all chemical compounds in microbes include carbon of some sort, whether these are fats, proteins, lipids, or carbohydrates (Rittmann, 2008).

2.4.1. Growth in Submerged Fermentation (SMF)

In the availability of appropriate media and the cultivation atmosphere, microbe enlargement is characterized as growth in all chemical components. For yeast and bacteria, cell number or cell mass growth may be quantified like a double of cell amount per unit time or a double of biomass per unit time for filament microorganisms like fungus. Upon injection of a sterile solution of nutrients with organisms and growth under physical conditions, four distinct stages of development are seen, depicted in Figure 2.6; development takes place as a consequence of nutrient intake (Stewart and Fogler, 2001).

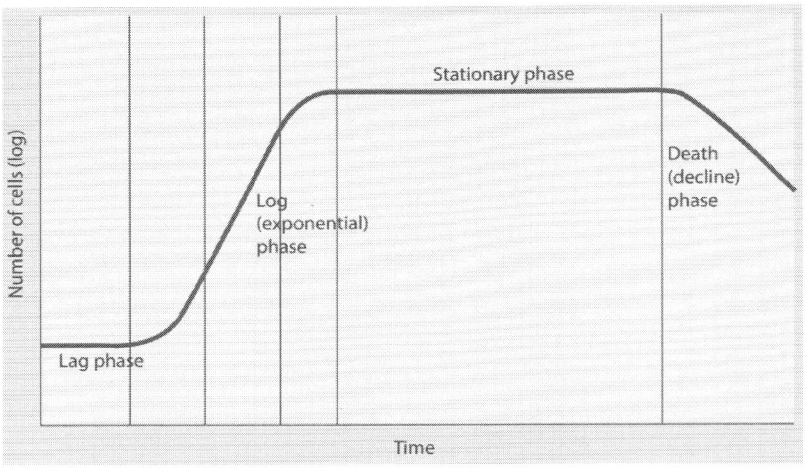

Figure 2.6. Four main microbial development stages.

Source: https://www.quora.com/What-are-the-phases-of-bacterial-growth-and-how-do-they-develop.

The cells absorb a portion of the substrate as well as necessary nutrients in the fermentation medium. First, the cells develop, and then they multiply. The growth pattern differs depending on whether the cells are molds or unicellular. Single-celled organisms, for example, that divide as they develop, will enhance the number of cells or the biomass (Tikhonenkov et al., 2020). They use more and more substrates as their numbers grow. The number of other sorts of cells (such as mold) would not grow, but their size and broth viscosity also increases. Cell counting is required for a variety of biological operations. The proportion may be determined by including the cells in a specific amount of culture. The proportion of microorganisms or cell counts (such as viruses, bacteria, and further parasites) in bodily fluids, for example, might disclose details regarding the progression of the disease and the immunity system of a person. In molecular biology studies, knowing the cell concentration allows you to alter the number of reagents and chemicals you use. The next sections go through the numerous methods for monitoring microbial growth (Singh et al., 2019).

2.4.1.1. Plate Count

The figure of constantly developing cells in a specimen of fluid may be evaluated using a feasible cell calculation (quantity of colonizing bacteria). A sterilized pipette is used to insert a predetermined volume of inoculum (usually one millimeter) from a sample in the center of a sterile Petri dish (consisting of nutritional media) in this approach (Figure 2.7). On a nutritional media, the bacteria develop in a colony (a group of cells that grow by a solitary bacterium through asexual reproduction). Every colony that develops after the sample is plated shows a spore or single cell in the original sample. The naked eye can see the colony, and the number of colonies on a plate may be calculated (colony count multiplied by the dilution factor) (Sanders, 2012).

2.4.1.2. Serial Dilution

A sequence of consecutive dilutions of a material in solution is known as a serial dilution. Each dilution would lower the sample's concentration by a certain amount, bringing it closer to a more acceptable concentration. The 1st step in producing a series of dilutions is to pour a precise quantity (typically one millimeter) of stock inside an identified volume (usually 9 millimeters) of purified water to produce a 10-fold (0.1 M) series of dilutions (Figure 2.8). The series of 10-fold dilutions might be 0.01 M, 0.001 M, 0.0001 M, and so forth. The aspect of dilution is relatively stable at each stage, resultant in a logarithmic geometric growth of the concentration (Detty and Merkel, 1990).

A ten-fold dilution decreases a solution or suspension's concentration by an aspect of 10 of its original concentration. The term "10-fold serial dilution" refers to a series of tenfold dilutions. They are performed in tiny sterilized test tubes, which are formed of glass and have fitting covers to reduce the possibility of contagion through the dilution.

The factor of dilution is a quantity that increases the final concentration by the actual concentration (cells/ml), lowering the stock solution's final concentration shown in Figure 2.9.

2.4.1.3. Most Probable Number Method

The Most feasible Number method involves duplicating the development of fluid broth in 10-fold or twice (10x or 2x) concentrations to find out the sum of feasible microorganisms in a specimen (Roos et al., 1995).

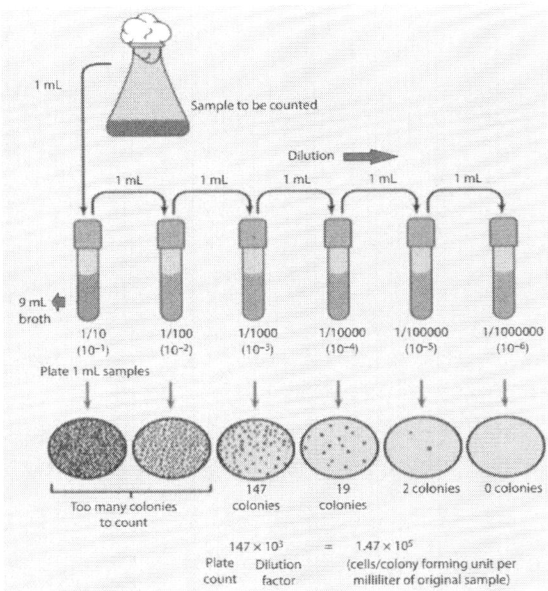

Figure 2.7. Count of the plate of live bacteria that form a colony.

Source: https://link.springer.com/book/10.1007/978-3-030-16230-6.

MPN examination is a statistical technique that relies upon the arbitrary dispersal of a microorganism's unit number in a known specimen. Calculated amounts of fluid are introduced to a sequence of tubes holding a fluid meter development media in this approach. Single or multiple indicators of bacteria causes the development and a distinct color alteration in the medium. Those bacteria that get a water inoculum without indicator do not see a color change. It is extensively utilized to guesstimate microbe's populations in the earth, waterways, and farming goods, and it is particularly effective when there is particle material in the specimen that hinders the conventional count of the plate by the methods of enumeration (Huang et al., 2018).

2.4.1.4. Optical Density

When measuring the concentration of growing bacteria, an optical density or spectrophotometer is generally used. Lambert-Beer's law is the foundation of this approach.

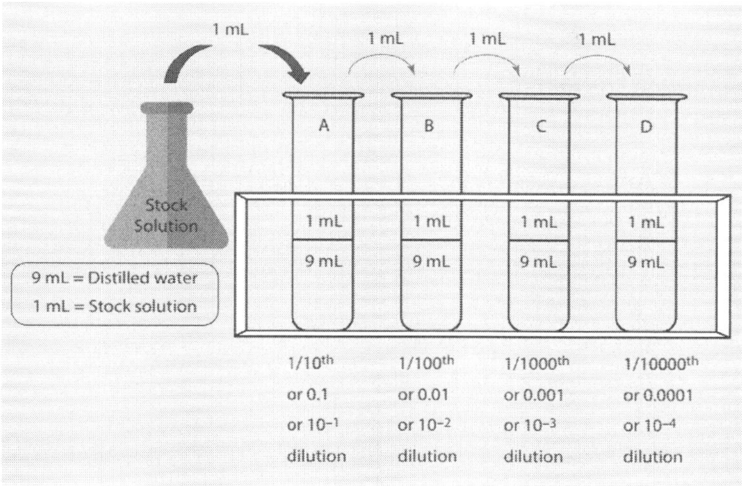

Figure 2.8. Series of in order dilution of store solution.

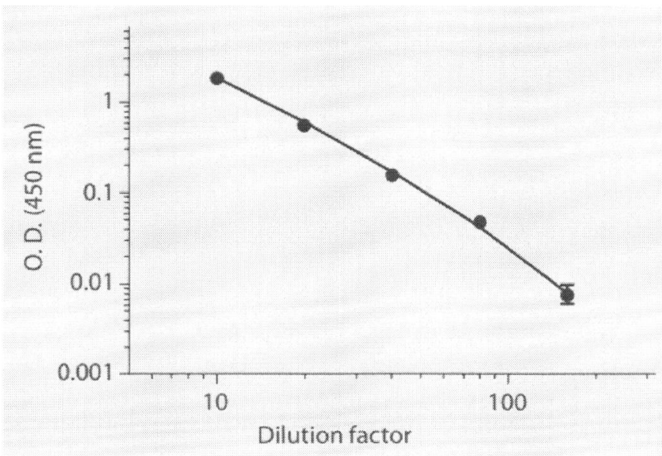

Figure 2.9. Graph showing optical density (O.D.) with the factor of dilution.

The particular array that is precise to the kind of the cell, optical density is directly related to the biomass of the cell suspension. The light intensity is absorbed and scattered by turbid cell concentrations or cell suspensions

(Hielscher et al., 1997). The turbidity rises in proportion to the concentration of the cells. The absorbance of the cell culture in a clear cuvette is evaluated in comparison to the medium alone (Figure 2.10).

Here is how the Lambert and Beer Law works:

"When a beam of monochromatic light of initial intensity (I_0) travels via a solution in a clear vessel, some of the light is absorbed, which result in a transmitted light intensity "I" that is lesser than " I_0." Light intensity has reduced due to scattering via particles in the solution and refraction at the interface (Crosby and Demas, 1971)."

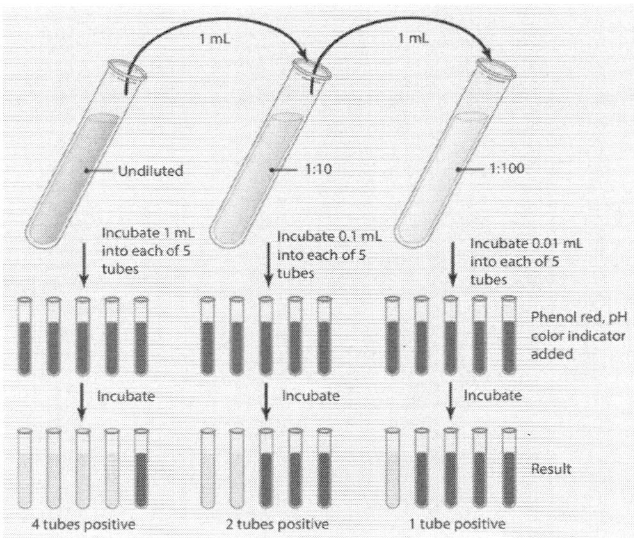

Figure 2.10. MPN method of microorganisms per volume of sample.

An electronic change absorbance is determined by two external factors:

- The absorbance is directly related to the light path length (l), which is the same as the cuvette's breadth.

Absorbance (A) ∝ light path (l)

- The absorbance of the mixture of the specimen utilized in the experimentation is directly related to the concentration (C).

Absorbance (A) ∝ concentration (C)

When the two relationships are combined:

$A = \varepsilon Cl$

By adding a constant, this proportionality may be transformed into an equation.

The Beer-Lambert law (shown in Figure 2.11) is the formula, and the constant ε is termed as moles extinction, and it is a quantification of the possibility of the electronic change. The electronic change is more likely when the molar absorptivity is high (Swinehart, 1962).

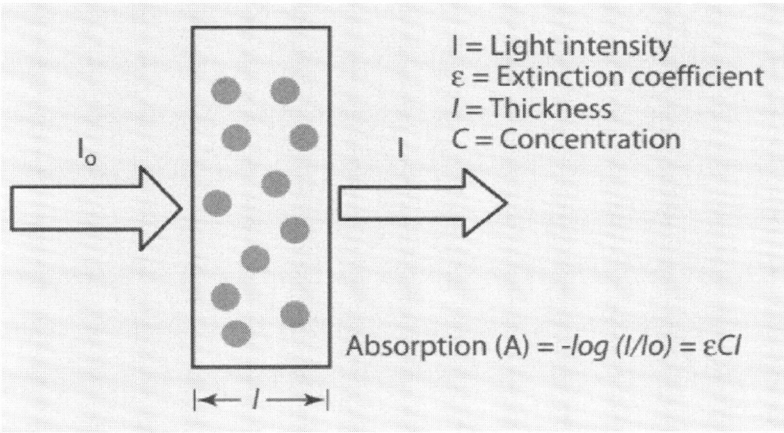

Figure 2.11. The Beer and Lambert law signifies A = εCl [where; A is the absorbance; constant ε is called molar extinction; l is the light path; C is the solution concentration, which is equivalent to the breadth of the cuvette].

Source: https://link.springer.com/book/10.1007/978-3-030-16230-6.

The composition of the solution of the specimen is calculated in mole per liter, and the pathway of light is measured in centimeters, by UV spectroscopy. Because absorbance has no units, molar absorptivity is measured in Liter/mole/centimeter.

Turbidimetry is an entrenched technique for studying the growth of bacterial (shown in Figure 2.12) because measurements of optical density allow for real-time monitoring of the growth of the bacterial population. The size distribution of cells is measured using Coulter counter numbers or electronic counting chambers. The assessment of a large numeral of bacteria in transparent fluid medium and soup are used to determine turbidity (Konstantinov et al., 1994).

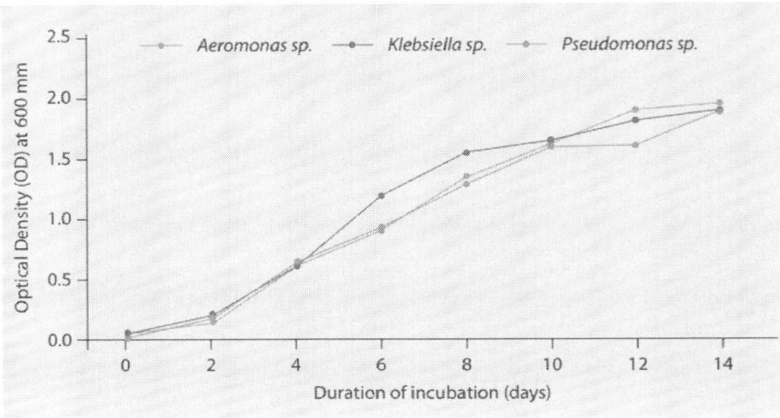

Figure 2.12. The method turbidimetry (optical density with the incubation period) of the growth of bacteria (Aeromonas sp., Klebsiella sp., Pseudomonas sp.).

Source: https://link.springer.com/book/10.1007/978-3-030-16230-6.

2.4.1.5. Kinetic Models for Cell Growth

Malthus' law widely describes how the development rate of cells on every stage in the development stage is proportionate to the numeral of cells present at the moment (Zakaria et al., 2014).

$$\frac{dx}{dt} = \mu x, \qquad x = x_0 \quad at\ t = 0 \tag{1}$$

where; μ is the constant of proportional identified as precise development rate in per hour; x is the mass of the cells per unit volume; and time in hours denoted by t.

By integration, we obtain:

$$ln\frac{x}{x_o} = \mu t \ \text{ or } \ x = x_o e^{\mu t} \tag{2}$$

$$ln\,x = ln\,x_o + \mu t \tag{3}$$

Eqn. (2) gives the time necessary to double the microbial mass. On a semi-logarithm plot of In x vs time, exponential development is represented by a straight line.

$$\tau_d = ln\frac{2}{\mu} = \frac{0.693}{\mu}$$
(4)

where; τ_d is the cell mass double time (Satyanarayana et al., 2003).

2.4.1.6. Monod Model

The development stage goes through several stages, including sluggish growth, strong growth, and ultimately termination. In other terms, the particular development rate is affected by the limiting substrate's residual concentration. Monod (1949) provided an empirical equation that explains the relationship effectively (Monod, 1949).

$$\mu = \frac{\mu_m S}{k_s + S}$$
(5)

where; μm is the highest growth rate that may be achieved and the limited concentration of the substrate is ks when the precise development rate is ½ of the maximum precise development rate, for example, μ = μm/2.

Substituting μm/2 for μ in Eqn. (5) demonstrates this, e.g.:

$$\frac{\mu_m}{2} = \frac{\mu_m S}{k_s + S}$$
(6)

By canceling μ_m on b/s and arrangement, $k_s = S$.

In Eqn. (5), if $S >> k_s$, the denominator can be compared to just S.

$$\mu = \frac{\mu_m S}{k_s + S} = \frac{\mu_m S}{S} = \mu_m$$
(7)

In other terms, Eqn. (7) shows the percentage in μ_m as a comparison to μ, when the value of S is large.

2.4.2. Growth in Solid-State Fermentation (SSF)

Due to the difficulty of extracting the biomass of microbes from the substance, direct measurement of biomass in SSF is extremely challenging. Because fungus hyphae penetrate and attach firmly to the substance, this is particularly factual for SSF procedures connecting fungus. The absolute value of microbial biomass, on the other side, is significant for calculating yields and rates of growth. The microbial population (colonization units) of yeasts and bacteria, on the other side, may be determined by sequential dilution of the fermented mash and discharge plating on a particular culture medium. The subsequent techniques for biomass measurement of fungus (as well as applied to microbes) in SSF have been described by Desgranges et al. (1991); and Terebiznik and Pilosof (1999). Biomass metabolic measurement is addressed in subsections.

2.4.2.1. Respiratory Metabolism

The metabolic mechanisms through which aerobic bacteria obtain the majority of their energy for development are CO_2 evolution and oxygen (O_2) consumption as a result of respiration. Carbon molecules in the substrate are digested, and CO_2 and microbial biomass are produced (Vries et al., 1974). During growth, CO_2 production makes the weight of fermenting substrate drop, and the quantity of weight loss may be linked to the quantity of growth. When combined with the usage of a correlation model, measuring oxygen consumption or CO_2 evolution becomes more powerful. Because continuous online data may be produced when both monitoring and computing equipment are present, such correlation models give a strong way of biomass estimation. The capability to check the coefficient of respiratory to achieve optimum substance oxidation, the capability to include automatic criticism control over the rate of aeration, and the anti-destructive nature of the measuring system are all benefits of using infrared and paramagnetic analyzers to observe concentrations of the effluent gas (Lanza et al., 2011).

2.4.2.2. Manufacturing of Basic Metabolites

A developing cell would usually utilize its basic metabolites to grow rather than exporting them to the media. Moreover, basic metabolite synthesis in processes of fermentation, such as enzymes, nucleotides, amino acids, and natural acids is frequently growth-dependent. Organic acid and mycelial growth and production have been found to have a strong link, which may be evaluated using pH calculation or a posteriori connected using HPLC examination on extracts. Sauer et al. (2008) established a tight relationship among fungus protein (biomass) and organic acids (fumaric, citric, acetic, or lactic) in Rhizopus. Though, because the product ratio produced per unit mass of the cell is not always the same, the connection with the cell mass is weak.

2.4.2.3. Biomass Components

- **Protein Content:** Protein is the easily calculated component of biomass. Because it detects peptide bonds, the *biuret technique is simple and repeatable, although it is not extremely sensitive.

Nonprotein nitrogen can cause errors when multiplying $Kjeldahl nitrogen by 6.25 to get crude protein. The Folin and Ciocalteu technique (**the technique of Lowry's) is very responsive, although it might provide an incorrect value if the reference and samples protein are identical in composition (Jernejc et al., 1986).

- **Nucleic Acid:** These techniques dependent on RNA/DNA identification are just effective when the substrate contains minimal nucleic acid and no interfering substances.
- **Glucosamine:** This technique is a good approach for estimating biomass fungus in SSF. The existence of chitin in the cell walls of several fungi is used in this approach. Chitin is made up of N-acetylglucosamine polymers. The formation of complicated agricultural substances including glucosamine in glycol-proteins may interfere with this technique (Puccia et al., 2013).
- **Ergosterol:** In fungus, ergosterol is the most abundant sterol. To determine Agaricus bisporus growth, glucosamine measurement is compared to ergosterol estimation.

2.5. FACTORS AFFECTING MICROBIAL GROWTH IN FERMENTATION

Microbes' development is affected through a variety of exterior and internal variables, including: temperature, pH, medium composition, and so on.

2.5.1. Temperature

Microorganisms require the right temperature to flourish. Growth happens slowly at temperatures below the optimal, resulting in a lower rate of cellular formation (shown in Figure 2.13). Based on their optimal growing temperature, bacteria may be classified into the following categories:

1. **Thermophiles:** These are microorganisms that thrive in hot environments (such as Thermococcus litoralis, Thermus aquaticus, and Bacillus stearothermophilus). The ideal temperature for growth is between 45 and 70°C, which may be observed in compost heaps and hot springs. Bacteria that thrive at extremely high temperatures are known as hyperthermophiles (such as genus Sulfolobus) (Harris et al., 1980). Their ideal temperature is between 70 and 110°C. They are generally Archaea members and

may be seen growing around hydrothermal vents at depths in the ocean (shown in Figure 2.14(b and d)).

Bacteria that flourish at medium temperatures are known as Mesophiles (such as Thiobacillus Novellus, Pseudomonas maltophilia, Streptococcus pneumonia, Clostridium kluyveri, Streptococcus pyrogens, etc.). Their ideal growing temperature is between 25 and 45°C. The majority of bacteria, including ordinary microorganisms that survive inside and outside of the body and soil bacteria, are mesophilic (Castenholz, 1969).

2. **Psychrophiles:** These are microbes that survive in cold environments (such as Psychrobacter sp., Arthrobacter sp.). Their ideal growing temperature is between -5 and 15°C. They are most commonly found in the Antarctic areas and Arctic (shown in Figure 2.14(a and c)) as well as in glacier-fed streams. In the fermentation process, these bacteria have a minor role.

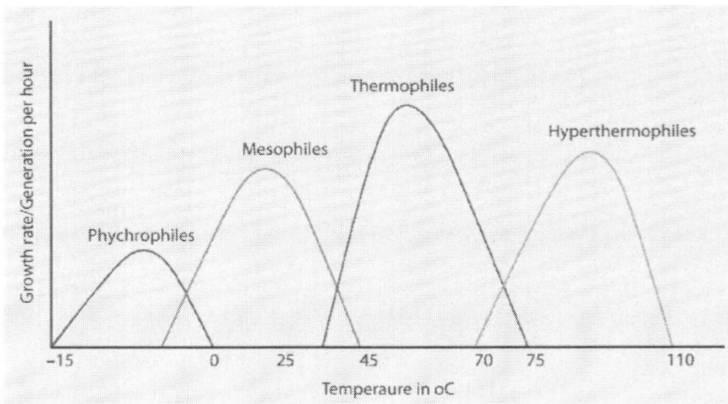

Figure 2.13. Development of bacteria (temperature vs. development rate in degree Celsius) based on temperature (mesophiles, thermophiles, psychrophile, and hyperthermophiles).

Source: https://courses.lumenlearning.com/microbiology/chapter/temperature-and-microbial-growth/.

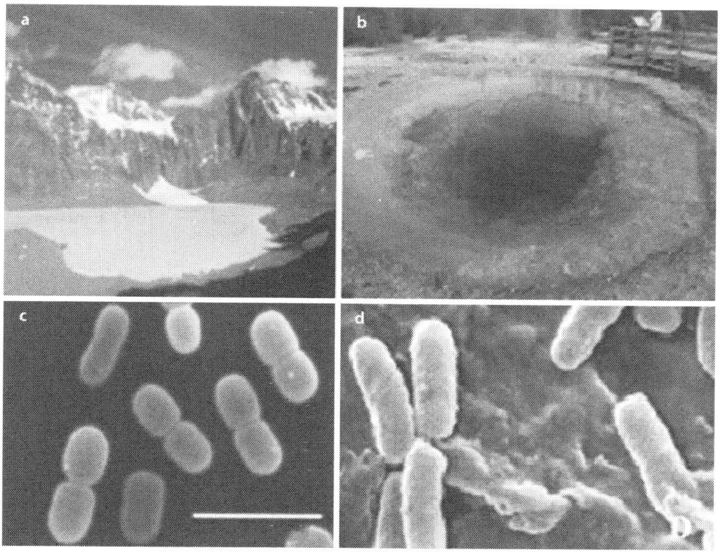

Figure 2.14. Archae members observed near hydrothermal vents of great depths in the ocean (b, d) and Antarctic and Arctic areas (a, c) in the watercourse supplied by glaciers.

Source: https://link.springer.com/chapter/10.1007/978-3-030-16230-6_1.

2.5.2. pH

The pH range around 5 and 7 is ideal for most bacteria (fungi and yeast develop in the conditions of acid). Based on respective optimal pH needs, microorganisms may be classified into one of the following categories: Neutrophils grow in the pH range of 5 to 8. Acidophiles develop in environments with a pH under 5.5. At a pH of above 8.5, alkaliphiles grow (Sarethy et al., 2011).

2.5.3. Oxygen

For aerobic organisms, oxygen (aeration) is a critical element that influences their growth rate. Many bacteria require significant quantities of dissolved oxygen to develop optimally. Because oxygen is only slightly soluble in water (8.4 mg/L at 25°C), it must be continually supplied to a developing culture (usually in the form of disinfected air). The agitator is utilized to split apart the froth and blend the contents in the process of fermentation. The rate of oxygen transfer to the cells is poor and inadequate to fulfill oxygen needs if the flow of air is insufficient or the bubbles of air are too big (Figure 2.15) (Marks, 2003).

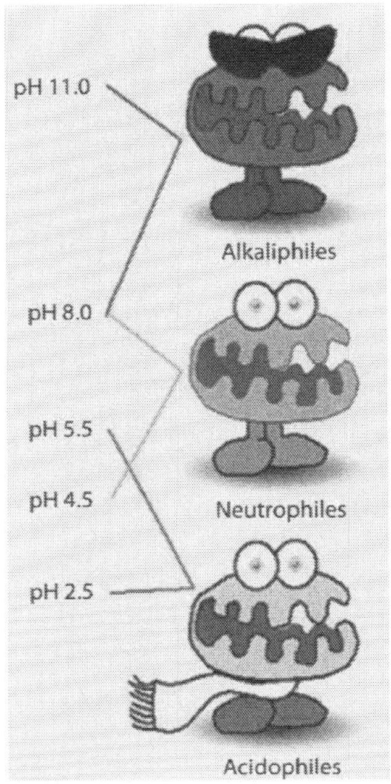

Figure 2.15. Microbes developing in the range of various pH.

Source: https://link.springer.com/chapter/10.1007/978-3-030-16230-6_1.

2.5.4. Constituents of Growth Medium

2.5.4.1. Macro- and Microelements

For metabolism and development, all bacteria require particular nutrients. Greater than 4 to 10 M concentrations are required for macroelements. Major elements include nitrogen, carbon, hydrogen, oxygen, phosphorus, sulfur, K^+ and $Mg^{2+.}$ Compounds of Carbon in the growth media are important sources of carbon and energy for cells. Microelements, on the other hand, must be present in concentrations less than 4 to 10 M. Micro-elements include elements like Ca^{2+}, Zn^{2+}, Mn^{2+}, Mo^{2+}, Na^+, Cu^{2+}, vitamins, metabolic precursors, and hormones of development (Altomare et al., 1999).

2.5.4.2. C:N Ratio of Growth Medium

The carbon to nitrogen ratio (C by N ratio) is the mass of carbon in a substrate divided by the amount of nitrogen. Microorganisms degrading organic materials with a greater C: N ratio confront an excess of Carbon in comparison to Nitrogen, whereas microbes with an inferior C to N ratio confront a deficiency of Carbon in comparison to Nitrogen (Grunert et al., 2016). Corn syrup, starch, waste sulfite liquor, and molasses are the most popular sources of carbon in the fermentation industries. Glucose, Fructose, and Sucrose are the majorly used sources of carbon in laboratory fermentation. Methanol, ethanol, and methane are all inexpensive carbon sources that may be used in fermentation. The majority of microorganisms utilized in the industry may use both organic and inorganic nitrogen sources. Ammonium salts, ammonia gas, and nitrates are all forms of inorganic nitrogen. Ammonia has been utilized as a primary source of nitrogen in a specified medium and for pH control as well (Mulay and Khale, 2011). Ammonium salts, such as ammonium sulfate, tend to make the growth media acidic. However, when they are digested, ammonium nitrates typically produce an alkaline drift (providing a basic environment).

2.5.4.3. Growth Factors

Some bacteria may be unable to generate a complete set of cell components, and requiring the use of growth factors, which are preformed compounds. Some metabolites' growth and production are stimulated by growth factors. The most important growth factors are vitamins, amino acids, and hormones. Riboflavin (B_2), thiamine (B_1), biotin, pyridoxine (B_6), folic acid, cyanocobalamin (B_{12}), lipoic acid, vitamin K, and p-aminobenzoic acid are some of the most widely utilized vitamins. Vitamins are the most frequent growth factors required, although particular amino acids, sterols, and fatty acids can also be needed (Phillips, 1984). All or some of the amino acids might have to be supplied outside, depending on the organism. Certain organisms required tiny amounts of certain fatty acids, like oleic acids and sterols. Several natural medium formulations include all or part of the necessary factor of growth. Biotin concentrations in the medium must be kept to a minimum in the procedures that produce glutamic acid. Some strains of production can also need thiamine (B_1).

2.6. CULTURES OF STARTER

2.6.1. Concept of the Cultures of Starter

A culture of starter is a preparation that contains a large number of desirable microorganisms and is used to speed up the process of fermentation. The preparations can comprise inevitable leftovers through the culture substance as well as others (such as non-freeze or non-oxidant chemicals) that help the microorganisms maintain their strength and technical functioning. After being adapted to the substrate, a typical starter allows for better control of the process of fermentation and prediction of the end yield (Santer, 2010). Three types of cultures starter are multi-strain mixed cultures which contain different strains from various classes; cultures of single-strain that have just a single strain of a class; and cultures of multi-strain that have several strains of a single class (Vogel et al., 2011).

2.6.2. Background and Succeeding Growth of Culture of the Starter

Microbes are found in raw substrates because they are inherently omnipresent. The concept of impulsive fermentation was born from this. Backslapping was a key technical phenomenon in spontaneous fermentation, which included vaccinating the raw matters with a tiny amount of a formerly flourishing fermentation. As a result of the best-adapted strains' dominance, backslapping occurred. This method is still employed in the manufacture of drinks and foods where the specific knowledge of the microbial role and population and ecology is unknown. This is also a cost-effective and dependable technique for fermenting foods (Solieri et al., 2013).

2.6.2.1. Recombinant Starter Culture

Genetic modification and recombinant technology provide a wide range of options for incorporating new characteristics into organisms, with favorable implications for the pharmaceutical and food sectors. The recombinant Lactococcus is a starter strain for producing peptides through Lactobacillus helveticus 53 by 7 to be used as an extra enzyme source while ripening of cheese (Joutsjoki et al., 2002). EPS (exopolysaccharide) synthesis can be increased by genetically engineering galactose metabolism in Streptococcus thermophilus strains. The addition of bio-thickeners to fat-free yogurts can be replaced by in situ synthesis of EPS by cultures of starter (Robitaille et al., 2009). Both Lactobacillus (Lactobacillus casei and Lactobacillus Plantarum

NC8) have been described as suitable appearance hosts for recombinant enzymes reported by Bohmer et al. (2012).

2.7. FACETS OF FERMENTATION

Humans utilized the process of fermentation for different reasons, counting bioremediation of organic wastes to the food valorization, industrial uses to agriculture and food, and agricultural wastes, among others.

2.7.1. Industrial Fermentation

The purposeful fermentation use by microorganisms, for example, microbes, and fungus to create an item beneficial to human beings is known as industrial fermentation.

2.7.1.1. Production of Biomass

Lactobacillus, singled cell protein (SCP), the yeast of bakery, and other biomass or microbial cells are occasionally the desired result of fermentation. Single-cell protein stands for single-celled microorganisms that are edible. Bio-mass (protein extort) by mixed or pure algae cultures, fungus, yeasts can be utilized as a substitute for meals excess protein, and it is safe to eat for humans and also safe to feed animals (Kawaguchi et al., 2001). The yeast of bakery (Saccharomyces cerevisiae) is a leavening agent in bakery products and baking bread that transforms fermentable sugars in the bread into ethanol and CO_2.

2.7.1.2. Production of Extracellular Metabolites

Basic metabolites (such as ethanol, lysine, vitamins, citric acid, polysaccharides, and glutamic acid) and secondary metabolites (gibberellins, lovastatin, cyclosporin-A, and penicillin) are the two types of metabolites (described above).

2.7.1.3. Production of Intracellular Components

Amylase, catalase, pectinase, protease, cellulase, glucose isomerase, lipase, hemicellulase, streptokinase, lactase, and other microbial enzymes are among the intracellular components. Furthermore, recombinant proteins made in the industries, like hepatitis B vaccination, interferon, granulocyte colony-stimulating factors, and insulin play a significant role in the health sector.

2.7.2. Food Fermentation

Food fermentation is a 1,000-year-old technique for making curds, bread, dosa, idli, cheese, and other foods. Although, a similar mechanism is used to convert sugars into ethanol-producing alcoholic beverages like beer, cider, and alcohol as well as to preserve sour foods like yogurt and sauerkraut through the formation of lactic acids (LAs). Grain, beans, fruit, vegetables, dairy products, honey, meat, fish, or tea can all be used to make more specialized fermented meals (Manavalan et al., 2021).

2.7.3. Biofuels

Fermentation For the manufacturing of butanol (such as butyl alcohol) and ethanol fuel (such as ethyl alcohol) fermentation is the primary source of ethanol. Yeast fermenting common crops including potato, sugarcane, corn, and cassava are utilized to generate ethanol, which is then processed into biofuel additives for gasoline or motor fuel.

2.7.4. Sewage Treatment

The process of eliminating pollutants from municipal wastewater is known as sewage treatment (or "wastewater treatment") (contains largely domestic sewage and also certain wastewater of industries). Sewage is processed by bacteria-secreted enzymes during the sewage treatment process.

2.7.5. Feeds of Animal

A broad range of Fermented farming desecrated products may be utilized for animal feed, particularly ruminants. To enhance the composition of protein and enhance the digestibility of Vitro, fungi have been utilized to dissolve the wastes of cellulosic compounds.

2.8. REPRESENTATIVE METABOLITES PRODUCED THROUGH MICROBES

Metabolites of microbes have a wide range of chemical properties. Microbes may access the compounds that synthetic chemists may not access. While there have been over 25,000 microbial metabolites recorded in the scientific literature, only around 2% of them have ever been made widely available to the research community. The biological function of most metabolites has never been thoroughly explored since they contain only a limited amount in the research lab where they had been found (Firn and Jones, 2003).

2.8.1. Primary Metabolites

Primary metabolites are important compounds (such as vitamins, amino acids, solvents, organic acids, or nucleotides) in all living cells that are the intermediate or end product of metabolic pathways, or are transformed into coenzymes, or are the home of essential macromolecules.

2.8.1.1. Amino Acids

The worth of the amino acid industry is more than 6 billion dollars and is expanding at a rate of 5 to 10% each year (Burkovski and Krämer, 2002). Monosodium glutamate, for example, is a top amino acid in terms of volume and is an efficient flavor enhancer. Several species of the genus Brevibacterium and Corynebacterium (for example, Brevibacterium lactofermentum, Brevibacterium flavum, and Corynebacterium glutamicum) have been identified as glutamate producers.

2.8.1.2. Nucleotides and Nucleosides

The action of 2 purine ribonucleoside 5′-monophosphates (such as inosinic acid (5′IMP) and guanylic acid (5′-GMP), causes nucleoside or nucleotide fermentations. Three main processes take place: (i) the hydrolysis of yeast RNA and IMP by enzymatic deamination; (ii) Bacillus subtilis mutants phosphorylate and synthesize nucleosides (guanosine and inosine); and (iii) transition of guanine toward GMP (by utilizing whole cells of Brevibacterium ammoniagenes) and transition of direct fermentation of sugar toward IMP (utilizing mutants of Corynebacterium glutamicum). The intracellular GMP and AMP are restricted to the efficient storage of purine nucleotides and are largely clarified through the feed control of purine auxotrophs (ASS), for example, adenine neutral mutants are poor in degradation of nucleotide enzymes (for example, adenylosuccinate synthetase (ASS)), which leads to hypoxanthine or inosine storage due to the disintegration of the accumulated intracellular IMP (Table 2.6) (Demain, 2007).

Table 2.6. Representative Metabolites Produced Through Various Groups of Microbes

Primary Metabolites	References	Microorganism Involved	Description and Usage
Amino Acids			
Monosodium glutamate	Bukowski and Kramer (2002)	Brevibacterium flavum, Brevibacterium lactofermentum, and Corynebacterium glutamicum	Effective flavor enhancer
Nucleotides and Nucleosides			
Inosine	Miyagawa et al. (1986)	Bacillus subtilis	IMP dehydrogenase (35 grams per liters), auxotrophs
Guanylic acid (5′-GMP) and inosinic acid (5′-IMP)	Nakayama et al. (2007)	Corynebacterium glutamicum, Brevibacterium ammoniagenes, and Bacillus subtilis,	ASS (auxotrophs)
Cytidine	Asahi et al. (1996)	Bacillus subtilis	HSD (30 grams per liters), auxotrophs
Inosine monophosphate or guanosine	Miyagawa et al. (1986)	Bacillus subtilis	IMP gene (7 to 20 grams per liters), auxotrophs
Vitamins			
Vitamin B_{12} (cyanocobalamin)	Spalla et al. (2010)	Propionibacterium shermanii and Pseudomonas denitrificans	Animals and humans nutrition (206 milligrams per liters)
Riboflavin (vitamin B_2)	Demain (1977)	Eremothecium ashbyii and Ashbyagossypii	Animals and humans nutrition (20 grams per liters)
Vitamin C (L-ascorbic acid)	Saito et al. (1997)	Erwinia herbicola by Corynebacterium sp.	Potent antioxidant (130 grams per liters)
Biotin antimetabolites	Masuda et al. (1995)	Serratia marcescens	Nutrition of animals and humans (600 milligrams per liters)
Organic Acids			

Gluconic acid	Znad et al. (2004)	Aspergillus niger	150 grams per liters
Oxalic and gluconic acids	Kubicek and Rohr (1985)	Aspergillus niger	IDH (pH 3)
Citric acid	Anastassiadis et al. (2002)	Candida oleophilic	80% sugar (100 grams per liters)
Gluconic acid	Kubicek and Rohr (1985)	Aspergillus niger	GO (pH 1.7 to 2)
Citric acid	Anastassiadis et al. (2002)	Aspergillus niger	Batch or Fed-batch fermentation (6 to 10 days) and (150 to 180 grams per liters)
Isocitric acid	Deppenmeier et al. (2002)	Candida sp.	Aconitase (225 rams per liters)
Lactic acid	Znad et al. (2004)	Lactobacilli	100 grams per liters
Pyruvic acid	Li et al. (29)	Torulopsis glabrata	69 grams per liters
Vinegar	Fukaya et al. (1989)	Acetobacter polyoxogenes, Acetobacter acetic subsp. xylinum.	AldDH (68 to 97 grams per liters)
Succinic acid	Zeikus et al. (1999)	Actinobacillus succinogenes	110 grams per liters
Itaconic acid	Wilke and Verlop (2010)	Candida sp.	–
Alcohols			
Glycerol	Taherzadeh et al. (2002)	Candida glycerinogenes	130 grams per liters
Ethanol	Ingram et al. (1987)	Escherichia Coli	ADH II and PDC (46 grams per liters)
Mannitol	Lee et al. (2003)	Candida magnolia	213 grams per liters
Secondary Metabolites			
Antibiotics			
Cephalosporin-C	Jiang et al. (2004)	Acremonium chrysogenum	30 grams per liters
Penicillin	Jiang et al. (2004)	Penicillium chrysogenum	70 grams per liters
Antitumor Agents			

Camptothecin (CPT)	Amna et al. (2006)	Endophytic fungi	Used against cancer
Mitomycin-C, daunorubicin, bleomycin, etoposide, doxorubicin, and calicheamicin	Strobel et al. (1996)	Actinomycetes	Effective agent against cancer
Pharmacological Agents			
Cyclosporin-A	Omura and Crump (2004)	Tolypocladium nivenum	Immunosuppressants
Lovastatin	Stabb et al. (1994)	Aspergillus terreus	Cholesterol-lowering agents

Abbreviations: 5′-GMP (guanosine 5′-monophosphates); 5′-IMP (Inosine 5′-monophosphates); ASS (adenylosuccinate synthetase); IMP (inosine monophosphate); HSD (homoserine dehydrogenase); IDH (isocitrate dehydrogenase); GO (glucose oxidase); AldDH (aldehyde dehydrogenase); ADH (alcohol dehydrogenase); PDC (pyruvate decarboxylase: 5–20% acetic acid).

2.8.1.3. *Vitamins*

Vitamins and vitamin-like substances (such as vitamins C, vitamins B_{12}, linolenic acid, vitamins F, riboflavin, ergosterol, and β-carotene) are produced by a variety of microbial strains. Riboflavin (vitamin B_2) has been professionally manufactured for several years through both chemical synthesis and fermentation; although, fermentation is now the preferred method. For example, Ashbya gossypii and Eremothecium ashbyii are two yeast-like molds, were discovered to synthesize riboflavin in amounts of more than 20 g/L. Pseudomonas denitrificans and Propionibacterium shermanii is utilized in the industrial manufacturing of vitamin B_{12} (cyanocobalamin) (Mera and Escalante, 2010). The inclusion of betaine is required for the excess production of vitamin B_{12} (the unknown control mechanism). Although P. freudenreicheii may generate 206 milligrams per liter of vitamin B_{12}, it is no longer a prominent commercial organism producer.

2.8.1.4. *Organic Acids*

For the industrial synthesis of natural (organic) acids, microorganisms are extensively utilized (for example, LA malic acid, citric acid, succinic acid, pyruvic acid, gluconic acid, tartaric acid acetic acid (AA), and itaconic acid). Citric acid is a form of organic acid that is easily digestible, absorbed, and has less toxicity. As a result, it is widely utilized (approximately 15%) in the food and medicine industries (as an acidifying, flavor-enhancing agent, and antioxidant). The EMP (Embden-Meyerhof pathway) produces Citric acid, which is the 1st stage in the TCA (tricarboxylic acid) cycle. According to Harmsen et al. (1992), a large amount of manufacturing of Citric acid is likewise connected with a higher level of fructose 2,6-biphosphate.

Gluconacetobacter sp. and Acetobacter sp. is the best bacteria for making vinegar. Vinegar, on the other hand, has roughly 5 to 20% AA. Fermentation has significantly reduced the need for LA to be synthesized chemically. The LA is made in the presence of less oxygen, with a carbohydrate yield of 95% (w/w), a concentration of over 100 g/L, and productivity of over 2 g/L. This is similar to methods that use LA bacteria. Lactobacilli create a mixture of isomers, while Rhizopus only produces LA. Rhizopus oryzae is the best candidate for development as it produces solely stereochemically pure LA. Itaconic acid is synthesized by the selective fungus (such as Candida sp.) (Peffer et al., 2013). It is employed as a comonomer in synthetic fibers and resins. Whereas microbial techniques for the other acids exist, they have yet to be exploited on a big scale. Succinic acid may be manufactured at 110 g/L from the rumen bacteria Actinobacillus succinogenes. Torulopsis glabrata produces pyruvic acid at a rate of 69 grams per liter in 56 hours with a yield of 0.62 g/g glucose.

2.8.1.5. Alcohols

Ethyl alcohol (C_2H_5OH) is a primary metabolite manufactured by the polysaccharide fermentation and carbohydrate that may be depolymerized to produce sugar by fermentation. For such fermentation, yeasts are preferable, although the species utilized based on the substance. For hexose fermentation, Saccharomyces cerevisiae is utilized, while Kluyveromyces fragilis or Candida sp. can be used if the substrate is lactose or pentose, respectively. Ethanol is synthesized in Brazil at a rate of 12.5 billion liters per year from sugarcane and is utilized as a 25% fuel blend or as a pure fuel. Every year, 30 million tons of wine and 60 million tons of beer are synthesized using beverage ethanol. Chemical synthesis of glycerol from petroleum feedstocks is the most common method, however, improved fermentation methods are available (Wang et al., 2001). Glycerol production by osmotolerant yeast strains (Candida glycerinogenes) may reach 130 grams per liter. Glycerol is synthesized in large quantities like a by-product of the oil and fat industries, as well as manufacture by propylene and, to a lesser degree, by fermentation of glucose utilizing Saccharomyces cerevisiae.

2.8.2. Secondary Metabolites

Secondary metabolites are substances synthesized late in the development cycle (idiophase) through strains of some microbial species (mostly fungi and actinomycetes). When growth is restricted due to the depletion of one main food source, secondary metabolites are produced (such as nitrogen, phosphate, and carbon). For instance, Penicillin production through Penicillium chrysogenum begins during glucose in the medium of culture is depleted and the fungus switches to lactose, a less quickly used glucose (Thomas et al., 2013). Secondary metabolites are distinguished by the fact that they are typically not synthesized during the fast development phase (such as trophophase), but rather during a later production stage (such as idiophase). There is a comparison/difference among secondary and main metabolites as shown in Table 2.7 and Figure 2.16. Secondary metabolites are critical for human health and nutrition and have a wide range of biological functions. For example, pesticides, antibiotics, animal, and plant growth factors, poisons, and other medications are all extremely important economically.

2.8.2.1. Antibiotics

The most popular and well-recognized secondary metabolites that are used as antimicrobial agents are known as antibiotics. Antibiotics and derivatives (for example, tetracyclines, β-lactam peptide antibiotics, aminoglycosides, macrolide polyketides, polyketides, etc.), transformed medicine in the 20th century, with over 160 antibiotics and derivatives in the market. Antibacterials represent 62% of the anti-infective market, whereas vaccines and sera immunoglobulins account for 13%, antifungals for 7%, anti-HIV antivirals for 12%, and non-HIV antivirals for 6%. Penicillin titers with Penicillium chrysogenum have reached 70 grams per liter, while Cephalosporin-C titers with Acremonium chrysogenum have surpassed 30 grams per liter. The titer of clavulanic acid produced through Streptomyces clavuligerus has been reported to be greater than 3 grams per liter (Petkovic et al., 2006).

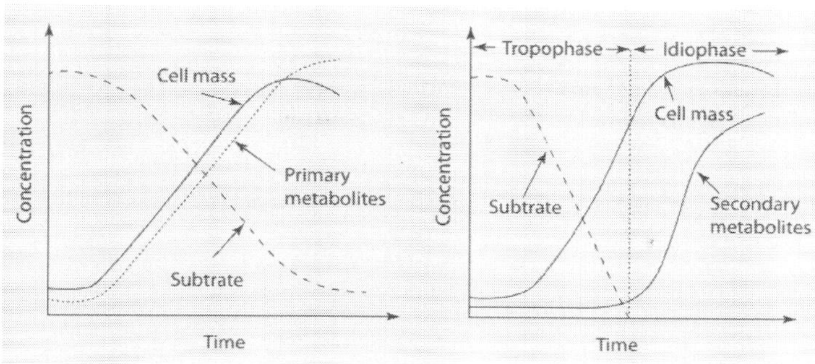

Figure 2.16. The comparison/difference among primary and secondary metabolites.

Source: https://link.springer.com/chapter/10.1007/978-3-030-16230-6_1.

Table 2.7. Comparison Among Primary Metabolites and Secondary Metabolites

Primary Metabolites	Secondary Metabolites
Tropophase is the development stage through which primary metabolites are generated.	The "idiophase" is the stage during which secondary metabolites are generated.
Produced during the cell's development phase	Produced during the cell's non-growth phase (close to completion).
Large amounts accumulated	Very small amount accumulated

2.8.2.2. Antitumor Agents

The major supply of anticancerous medicines are via microorganisms since the Actinomycins origination/discovery by Woodruff and Waksman (1941) and use of Actinomycin-D in children to control Wilms tumors. Bleomycin, mitomycin-C, doxorubicin, daunorubicin, calicheamicin, and etoposide are all-important microbial compounds synthesized by Actinomycetes. Taxol (paclitaxel) is a highly effective substance against breast cancer and ovarian compound synthesized by endophytic fungus. Camptothecin (CPT) is an advanced monoterpene indole alkaloid and is a dynamic anti-type I-deoxyribonucleic acid topoisomerase, which is another plant product synthesized by some angiosperms. Irinotecan and topotecan, two water-soluble derivatives, are utilized to treat cancer. The endophytic fungus may also produce it.

REFERENCES

1. Acosta, A., Sammarco, P. W., & Duarte, L. F., (2001). Asexual reproduction in a zoanthid by fragmentation: The role of exogenous factors. Bulletin of Marine Science, 68(3), 363–381.
2. Ajeej, A., Thanikal, J. V., Narayanan, C. M., & Kumar, R. S., (2015). An overview of bioaugmentation of methane by anaerobic co-digestion of municipal sludge along with microalgae and waste paper. Renewable and Sustainable Energy Reviews, 50, 270–276.
3. Altomare, C., Norvell, W. A., Björkman, T. H. O. M. A. S., & Harman, G. E., (1999). Solubilization of phosphates and micronutrients by the plant-growth-promoting and biocontrol fungus Trichoderma harzianum Rifai 1295–22. Applied and Environmental Microbiology, 65(7), 2926–2933.
4. Amna, T., Puri, S. C., Verma, V., Sharma, J. P., Khajuria, R. K., Musarrat, J., & Qazi, G. N., (2006). Bioreactor studies on the endophytic fungus Entrophospora infrequent for the production of an anticancer alkaloid camptothecin. Canadian Journal of Microbiology, 52(3), 189–196.
5. Anastassiadis, S., Aivasidis, A., & Wandrey, C., (2002). Citric acid production by Candida strains under intracellular nitrogen limitation. Applied Microbiology and Biotechnology, 60(1), 81–87.
6. Andersen, R., Buechi, G., Kobbe, B., & Demain, A. L., (1977). Secalonic acids D and F are toxic metabolites of Aspergillus aculeatus. The Journal of Organic Chemistry, 42(2), 352–353.
7. Asahi, S., Tsunemi, Y., & Doi, M., (1996). Effects of homoserine dehydrogenase deficiency on production of cytidine by mutants of Bacillus subtilis. Bioscience, Biotechnology, and Biochemistry, 60(2), 353–354.
8. Babu, K. R., & Satyanarayana, T., (1996). Production of bacterial enzymes by solid-state fermentation. Journal of Scientific and Industrial Research, 55(5–6), 464–467.
9. Behera, S. S., & Ray, R. C., (2016). Solid-state fermentation for production of microbial cellulases: Recent advances and improvement strategies. International Journal of Biological Macromolecules, 86, 656–669.
10. Behera, S. S., Ray, R. C., Das, U., Panda, S. K., & Saranraj, P., (2019). Microorganisms in fermentation. Essentials in Fermentation Technology, 2(31), 1–39.

11. Blair, M. W., Izquierdo, P., Astudillo, C., & Grusak, M. A., (2013). A legume biofortification quandary: Variability and genetic control of seed coat micronutrient accumulation in common beans. Frontiers in Plant Science, 4, 275.
12. Böhmer, N., Lutz-Wahl, S., & Fischer, L., (2012). Recombinant production of hyperthermostable CelB from Pyrococcus furiosus in Lactobacillus sp. Applied Microbiology and Biotechnology, 96(4), 903–912.
13. Bosiacka, B., Myśliwy, M., & Bosiacki, M., (2020). Habitat conditions strongly affect macro- and microelement concentrations in Taraxacum microspecies growing on coastal meadows along a soil salinity gradient. Peer J., 8, e10233.
14. Boyle, R., (2017). Eukaryotic origins and the proterozoic earth system: A link between global scale glaciations and eukaryogenesis?. Earth-Science Reviews, 174, 22–38.
15. Braga, J. W. B., Trevizan, L. C., Nunes, L. C., Rufini, I. A., Santos, Jr. D., & Krug, F. J., (2010). Comparison of univariate and multivariate calibration for the determination of micronutrients in pellets of plant materials by laser-induced breakdown spectrometry. Spectrochimica Acta Part B: Atomic Spectroscopy, 65(1), 66–74.
16. Burkovski, A., & Krämer, R., (2002). Bacterial amino acid transport proteins: Occurrence, functions, and significance for biotechnological applications. Applied Microbiology and Biotechnology, 58(3), 265–274.
17. Castenholz, R. W., (1969). Thermophilic blue-green algae and the thermal environment. Bacteriological Reviews, 33(4), 476–504.
18. Cavalier-Smith, T., (1998). A revised six-kingdom system of life. Biological Reviews, 73(3), 203–266.
19. Crosby, G. A., & Demas, J. N., (1971). Measurement of photoluminescence quantum yields: Review. The Journal of Physical Chemistry, 75(8), 991–1024.
20. Dadd, R. H., (1973). Insect nutrition: Current developments and metabolic implications. Annual Review of Entomology, 18(1), 381–420.
21. Dallner, G., Siekevitz, P., & Palade, G. E., (1966). Biogenesis of endoplasmic reticulum membranes: I. Structural and chemical differentiation in developing rat hepatocyte. The Journal of Cell Biology, 30(1), 73–96.

22. De Pauw, N., De Leenheer, L., Verlet, H., & Dochy, M., (1979). Culturing of marine microscopic algae. International Council for the Exploration of the Sea: ICES, Mariculture Committee, F, 54, 185–195.
23. De Swaaf, M. E., Sijtsma, L., & Pronk, J. T., (2003). High-cell-density fed-batch cultivation of the docosahexaenoic acid-producing marine alga Crypthecodinium cohnii. Biotechnology and Bioengineering, 81(6), 666–672.
24. De Vries, F. P., Brunsting, A. H. M., & Van Laar, H. H., (1974). Products, requirements and efficiency of biosynthesis a quantitative approach. Journal of Theoretical Biology, 45(2), 339–377.
25. Demain, A. L., (2007). Reviews: The business of biotechnology. Industrial Biotechnology, 3(3), 269–283.
26. Deppenmeier, U., Hoffmeister, M., & Prust, C., (2002). Biochemistry and biotechnological applications of gluconobacter strains. Applied Microbiology and Biotechnology, 60(3), 233–242.
27. Desgranges, C., Vergoignan, C., Georges, M., & Durand, A., (1991). Biomass estimation in solid-state fermentation: I. Manual biochemical methods. Applied Microbiology and Biotechnology, 35(2), 200–205.
28. Detty, M. R., & Merkel, P. B., (1990). Chalcogenapyrylium dyes as potential photochemotherapeutic agents. Solution studies of heavy atom effects on triplet yields, quantum efficiencies of singlet oxygen generation, rates of reaction with singlet oxygen, and emission quantum yields. Journal of the American Chemical Society, 112(10), 3845–3855.
29. Fife, C. M., McCarroll, J. A., & Kavallaris, M., (2014). Movers and shakers: Cell cytoskeleton in cancer metastasis. British Journal of Pharmacology, 171(24), 5507–5523.
30. Firn, R. D., & Jones, C. G., (2003). Natural products: A simple model to explain chemical diversity. Natural Product Reports, 20(4), 382–391.
31. Floros, J. D., Newsome, R., Fisher, W., Barbosa-Cánovas, G. V., Chen, H., Dunne, C. P., & Ziegler, G. R., (2010). Feeding the world today and tomorrow: The importance of food science and technology: An IFT scientific review. Comprehensive Reviews in Food Science and Food Safety, 9(5), 572–599.
32. Fukaya, M., Tayama, K., Tamaki, T., Tagami, H., Okumura, H., Kawamura, Y., & Beppu, T., (1989). Cloning of the membrane-bound aldehyde dehydrogenase gene of acetobacter polyoxogenes

and improvement of acetic acid production by use of the cloned gene. Applied and Environmental Microbiology, 55(1), 171–176.

33. Gest, H., (2004). The discovery of microorganisms by Robert Hooke and Antoni Van Leeuwenhoek, fellows of the Royal Society. Notes and Records of the Royal Society of London, 58(2), 187–201.
34. Gomez-Flores, M., Nakhla, G., & Hafez, H., (2015). Microbial kinetics of Clostridium termitidis on cellobiose and glucose for biohydrogen production. Biotechnology Letters, 37(10), 1965–1971.
35. Grunert, O., Reheul, D., Van Labeke, M. C., Perneel, M., Hernandez-Sanabria, E., Vleminck, S. E., & Boon, N., (2016). Growing media constituents determine the microbial nitrogen conversions in organic growing media for horticulture. Microbial Biotechnology, 9(3), 389–399.
36. Harmsen, H. J., Kubicek-Pranz, E. M., Röhr, M., Visser, J., & Kubicek, C. P., (1992). Regulation of 6-phosphofructo-2-kinase from the citric-acid-accumulating fungus Aspergillus niger. Applied Microbiology and Biotechnology, 37(6), 784–788.
37. Harold, F. M., (1966). Inorganic polyphosphates in biology: Structure, metabolism, and function. Bacteriological Reviews, 30(4), 772–794.
38. Harris, J. I., Hocking, J. D., Runswick, M. J., Suzuki, K., & Walker, J. E., (1980). D-glyceraldehyde-3-phosphate dehydrogenase. The purification and characterization of the enzyme from the thermophiles Bacillus stearothermophilus and Thermus aquaticus. European Journal of Biochemistry, 108(2), 535–547.
39. Hielscher, A. H., Mourant, J. R., & Bigio, I. J., (1997). Influence of particle size and concentration on the diffuse backscattering of polarized light from tissue phantoms and biological cell suspensions. Applied Optics, 36(1), 125–135.
40. Holzapfel, W., (1997). Use of starter cultures in fermentation on a household scale. Food Control, 8(5–6), 241–258.
41. Huang, Z., Marra, F., Subbiah, J., & Wang, S., (2018). Computer simulation for improving radio frequency (RF) heating uniformity of food products: A review. Critical Reviews in Food Science and Nutrition, 58(6), 1033–1057.
42. Ingram, L. O., Conway, T., Clark, D. P., Sewell, G. W., & Preston, J. F., (1987). Genetic engineering of ethanol production in Escherichia coli. Applied and Environmental Microbiology, 53(10), 2420–2425.

43. Jemil, I., Jridi, M., Nasri, R., Ktari, N., Salem, R. B. S. B., Mehiri, M., & Nasri, M., (2014). Functional, antioxidant and antibacterial properties of protein hydrolysates prepared from fish meat fermented by Bacillus subtilis A26. Process Biochemistry, 49(6), 963–972.
44. Jernejc, K., Cimerma, A., & Perdih, A., (1986). Comparison of different methods for protein determination in Aspergillus niger mycelium. Applied Microbiology and Biotechnology, 23(6), 445–448.
45. Joshi, R. C., Yadav, S., Pathak, V. K., Malhotra, H. S., Khokhar, H. V. S., Parihar, A., & Dutta, M. K., (2021). A deep learning-based COVID-19 automatic diagnostic framework using chest X-ray images. Biocybernetics and Biomedical Engineering, 41(1), 239–254.
46. Joutsjoki, V., Luoma, S., Tamminen, M., Kilpi, M., Johansen, E., & Palva, A., (2002). Recombinant Lactococcus starters as a potential source of additional pectinolytic activity in cheese ripening. Journal of Applied Microbiology, 92(6), 1159–1166.
47. Kaksonen, A. H., Boxall, N. J., Gumulya, Y., Khaleque, H. N., Morris, C., Bohu, T., & Lakaniemi, A. M., (2018). Recent progress in biohydrometallurgy and microbial characterization. Hydrometallurgy, 180, 7–25.
48. Kar, S., & Ray, R. C., (2011). Purification, characterization, and application of thermostable exo-polygalacturonase from Streptomyces erumpens MTCC 7317. Journal of Food Biochemistry, 35(1), 133–147.
49. Kar, S., Ray, R. C., & Mohapatra, U. B., (2008). α-amylase production by Streptomyces erumpens MTCC 7317 in solid-state fermentation using response surface methodology (RSM). Polish Journal of Microbiology, 57(4), 289.
50. Kaul, S., Koo, H. L., Jenkins, J., Rizzo, M., Rooney, T., Tallon, L. J., & Somerville, M. C., (2000). Analysis of the genome sequence of the flowering plant Arabidopsis thaliana. Nature, 408(6814), 796–815.
51. Kawaguchi, H., Hashimoto, K., Hirata, K., & Miyamoto, K., (2001). H_2 production from algal biomass by a mixed culture of Rhodobium marinum A-501 and Lactobacillus amylovorus. Journal of Bioscience and Bioengineering, 91(3), 277–282.
52. Khan, S., & Scholey, J. M., (2018). Assembly, functions, and evolution of archaella, flagella and cilia. Current Biology, 28(6), R278-R292.
53. King, R. D., Whelan, K. E., Jones, F. M., Reiser, P. G., Bryant, C. H., Muggleton, S. H., & Oliver, S. G., (2004). Functional genomic

hypothesis generation and experimentation by a robot scientist. Nature, 427(6971), 247–252.

54. Konstantinov, K., Chuppa, S., Sajan, E., Tsai, Y., Yoon, S., & Golini, F., (1994). Real-time biomass-concentration monitoring in animal-cell cultures. Trends in Biotechnology, 12(8), 324–333.
55. Kubicek, C. P., Röhr, M., & Rehm, H. J., (1985). Citric acid fermentation. Critical Reviews in Biotechnology, 3(4), 331–373.
56. Kumar, G., Cho, S. K., Sivagurunathan, P., Anburajan, P., Mahapatra, D. M., Park, J. H., & Pugazhendhi, A., (2018). Insights into evolutionary trends in molecular biology tools in microbial screening for biohydrogen production through dark fermentation. International Journal of Hydrogen Energy, 43(43), 19885–19901.
57. Lanza, I. R., Bhagra, S., Nair, K. S., & Port, J. D., (2011). Measurement of human skeletal muscle oxidative capacity by 31P-MR spectroscopy: A cross-validation with in vitro measurements. Journal of Magnetic Resonance Imaging, 34(5), 1143–1150.
58. Lee, J. K., Song, J. Y., & Kim, S. Y., (2003). Controlling substrate concentration in fed-batch Candida magnoliae culture increases mannitol production. Biotechnology Progress, 19(3), 768–775.
59. Li, S., Chen, X., Liu, L., & Chen, J., (2016). Pyruvate production in Candida glabrata: Manipulation and optimization of physiological function. Critical Reviews in Biotechnology, 36(1), 1–10.
60. Manavalan, A., Qader, M., Wu, X., & Cao, S., (2021). Nrf2 activation by Morinda citrifolia L.(Noni) fruit juices. World Journal of Traditional Chinese Medicine, 7(2), 184.
61. Marks, D. M., (2003). Equipment design considerations for large scale cell culture. Cytotechnology, 42(1), 21–33.
62. Masuda, M., Takahashi, K., Sakurai, N., Yanagiya, K., Komatsubara, S., & Tosa, T., (1995). Further improvement of D-biotin production by a recombinant strain of Serratia marcescens. Process Biochemistry, 30(6), 553–562.
63. Matz, C., & Kjelleberg, S., (2005). Off the hook–how bacteria survive protozoan grazing. Trends in Microbiology, 13(7), 302–307.
64. Mazzanti, M., Bustamante, J. O., & Oberleithner, H., (2001). Electrical dimension of the nuclear envelope. Physiological Reviews, 81(1), 1–19.

65. Mera, P. E., & Escalante-Semerena, J. C., (2010). Multiple roles of ATP: Cob (I) alamin adenosyltransferases in the conversion of B 12 to coenzyme B 12. Applied Microbiology and Biotechnology, 88(1), 41–48.
66. Mienda, B. S., Idi, A., & Umar, A., (2011). Microbiological features of solid-state fermentation and its applications: An overview. Research in Biotechnology, 2(6), 2–9.
67. Miyagawa, K. I., Kimura, H., Nakahama, K., Kikuchi, M., Doi, M., Akiyama, S. I., & Nakao, Y., (1986). Cloning of the Bacillus subtilis IMP dehydrogenase gene and its application to increased production of guanosine. Bio/Technology, 4(3), 225–228.
68. Monod, J., (1949). The growth of bacterial cultures. Annual Review of Microbiology, 3(1), 371–394.
69. Mulay, S., & Khale, A., (2011). Asavarishtas through improved fermentation technology. International Journal of Pharmaceutical Sciences and Research, 2(6), 1421.
70. Navarre, W. W., & Schneewind, O., (1999). Surface proteins of gram-positive bacteria and mechanisms of their targeting to the cell wall envelope. Microbiology and Molecular Biology Reviews, 63(1), 174–229.
71. Nicolson, G. L., & Ash, M. E., (2017). Membrane lipid replacement for chronic illnesses, aging, and cancer using oral glycerol phospholipid formulations with fructooligosaccharides to restore phospholipid function in cellular membranes, organelles, cells, and tissues. Biochimica et Biophysica Acta (BBA)-Biomembranes, 1859(9), 1704–1724.
72. Oguntoyinbo, F. A., Tourlomousis, P., Gasson, M. J., & Narbad, A., (2011). Analysis of bacterial communities of traditional fermented West African cereal foods using culture-independent methods. International Journal of Food Microbiology, 145(1), 205–210.
73. Õmura, S., & Crump, A., (2004). The life and times of ivermectin: A success story. Nature Reviews Microbiology, 2(12), 984–989.
74. Oren, A., (2009). Systematics of archaea and bacteria. Biological Science Fundamentals and Systematics, 2 (3), 162.
75. Peffer, T. E., Bodzin, A. M., & Smith, J. D., (2013). The use of technology by nonformal environmental educators. The Journal of Environmental Education, 44(1), 16–37.

76. Petkovic, H., Cullum, J., Hranueli, D., Hunter, I. S., Perić-Concha, N., Pigac, J., & Long, P. F., (2006). Genetics of Streptomyces rimosus, the oxytetracycline producer. Microbiology and Molecular Biology Reviews, 70(3), 704–728.

77. Phillips, N. W., (1984). Role of different microbes and substrates as potential suppliers of specific, essential nutrients to marine detritivores. Bulletin of Marine Science, 35(3), 283–298.

78. Puccia, R. M., Vallejo, M. C., & Larissa, V. G., (2013). The cell wall of the dimorphic pathogenic species of Paracoccidioides. Hector Mora Montes.(Org.), 47–90.

79. Qian, P. Y., Xu, Y., & Fusetani, N., (2009). Natural products as antifouling compounds: recent progress and future perspectives. Biofouling, 26(2), 223–234.

80. Rengel, Z., Batten, G. D., & Crowley, D. D., (1999). Agronomic approaches for improving the micronutrient density in edible portions of field crops. Field Crops Research, 60(1–2), 27–40.

81. Ribes, J. A., Vanover-Sams, C. L., & Baker, D. J., (2000). Zygomycetes in human disease. Clinical Microbiology Reviews, 13(2), 236–301.

82. Rittmann, B. E., (2008). Opportunities for renewable bioenergy using microorganisms. Biotechnology and Bioengineering, 100(2), 203–212.

83. Robitaille, G., Tremblay, A., Moineau, S., St-Gelais, D., Vadeboncoeur, C., & Britten, M., (2009). Fat-free yogurt made using a galactose-positive exopolysaccharide-producing recombinant strain of Streptococcus thermophilus. Journal of Dairy Science, 92(2), 477–482.

84. Roos, D. S., Donald, R. G., Morrissette, N. S., & Moulton, A. L. C., (1995). Molecular tools for genetic dissection of the protozoan parasite Toxoplasma gondii. Methods in Cell Biology, 45, 27–63.

85. Saito, Y., Ishii, Y., Hayashi, H., Imao, Y., Akashi, T., Yoshikawa, K., & Shimomura, K., (1997). Cloning of genes coding for L-sorbose and L-sorbosone dehydrogenases from Gluconobacter oxydans and microbial production of 2-keto-L-gulonate, a precursor of L-ascorbic acid, in a recombinant G. oxydans strain. Applied and Environmental Microbiology, 63(2), 454–460.

86. Sanders, E. R., (2012). Aseptic laboratory techniques: Plating methods. JoVE (Journal of Visualized Experiments), 63, e3064.

87. Santer, M., (2010). Joseph Lister: First use of a bacterium as a 'model organism' to illustrate the cause of infectious disease of humans. Notes and Records of the Royal Society, 64(1), 59–65.
88. Santini, T. C., Kerr, J. L., & Warren, L. A., (2015). Microbially-driven strategies for bioremediation of bauxite residue. Journal of Hazardous Materials, 293, 131–157.
89. Santos, C., Fraga, M. E., Kozakiewicz, Z., & Lima, N., (2010). Fourier transform infrared as a powerful technique for the identification and characterization of filamentous fungi and yeasts. Research in Microbiology, 161(2), 168–175.
90. Sarethy, I. P., Saxena, Y., Kapoor, A., Sharma, M., Sharma, S. K., Gupta, V., & Gupta, S., (2011). Alkaliphilic bacteria: Applications in industrial biotechnology. Journal of Industrial Microbiology and Biotechnology, 38(7), 769.
91. Sato, S., & Harris, J. I., (1977). Superoxide dismutase from Thermus aquaticus: Isolation and characterization of manganese and apoenzymes. European Journal of Biochemistry, 73(2), 373–381.
92. Satyanarayana, A., Prabhakara, R. P. G., & Rao, D. G., (2003). Chemistry, processing, and toxicology of annatto (Bixa orellana L.). J. Food Sci. Technol., 40(2), 131–141.
93. Sauer, M., Porro, D., Mattanovich, D., & Branduardi, P., (2008). Microbial production of organic acids: expanding the markets. Trends in Biotechnology, 26(2), 100–108.
94. Sengun, I. Y., & Karabiyikli, S., (2011). Importance of acetic acid bacteria in food industry. Food Control, 22(5), 647–656.
95. Seufferheld, M. J., Kim, K. M., Whitfield, J., Valerio, A., & Caetano-Anollés, G., (2011). Evolution of vacuolar proton pyrophosphatase domains and volutin granules: Clues into the early evolutionary origin of the acidocalcisome. Biology Direct, 6(1), 1–15.
96. Shaktimay, K., Datta, T. K., & Ray, R. C., (2010). Optimization of thermostable α-amylase production by Streptomyces erumpens MTCC 7317 in solid-state fermentation using cassava fibrous residue. Brazilian Archives of Biology and Technology, 53(2), 301–309.
97. Shieh, Y. C., & Beuchat, L. R., (1982). Microbial changes in fermented peanut and soybean pastes containing kojis prepared using Aspergillus oryzae and Rhizopus oligosporus. Journal of Food Science, 47(2), 518–522.

98. Shirokikh, I. G., & Shirokikh, A. A., (2017). Biosynthetic potential of actinomycetes in brown forest soil on the eastern coast of the Aegean sea. Eurasian Soil Science, 50(11), 1311–1317.

99. Shun-Jin, J., Ya-Yong, Y., & Hui-Qing, W., (2004). Optimization of clavulanic acid fermentation. Chinese Journal of Antibiotics, 29(6), 335–337.

100. Singh, H., Tiwari, K., Tiwari, R., Pramanik, S. K., & Das, A., (2019). Small molecules as fluorescent probes for monitoring intracellular enzymatic transformations. Chemical Reviews, 119(22), 11718–11760.

101. Singleton, R., Nanjundaswamy, A. K., Mandyam, K., & Njiti, V., (2017). Fermentation optimization of macro-fungus Pleurotus Sajor-Caju on soymeal. Ferment. Technol., 6, 146.

102. Solieri, L., Dakal, T. C., & Giudici, P., (2013). Next-generation sequencing and its potential impact on food microbial genomics. Annals of Microbiology, 63(1), 21–37.

103. Soydemir, G., Keris-Sen, U. D., Sen, U., & Gurol, M. D., (2016). Biodiesel production potential of mixed microalgal culture grown in domestic wastewater. Bioprocess and Biosystems Engineering, 39(1), 45–51.

104. Stabb, E. V., Jacobson, L. M., & Handelsman, J. O., (1994). Zwittermicin A-producing strains of Bacillus cereus from diverse soils. Applied and Environmental Microbiology, 60(12), 4404–4412.

105. Stewart, T. L., & Fogler, H. S., (2001). Biomass plug development and propagation in porous media. Biotechnology and Bioengineering, 72(3), 353–363.

106. Stock, J. B., Rauch, B., & Roseman, S., (1977). Periplasmic space in Salmonella typhimurium and Escherichia coli. Journal of Biological Chemistry, 252(21), 7850–7861.

107. Strobel, G. A., Hess, W. M., Ford, E., Sidhu, R. S., & Yang, X., (1996). Taxol from fungal endophytes and the issue of biodiversity. Journal of Industrial Microbiology, 17(5), 417–423.

108. Strobel, G., (2018). The emergence of endophytic microbes and their biological promise. Journal of Fungi, 4(2), 57.

109. Swinehart, D. F., (1962). The beer-lambert law. Journal of Chemical Education, 39(7), 333.

110. Taherzadeh, M. J., Adler, L., & Lidén, G., (2002). Strategies for enhancing fermentative production of glycerol: A review. Enzyme and Microbial Technology, 31(1–2), 53–66.
111. Tan, D. X., Manchester, L. C., Terron, M. P., Flores, L. J., & Reiter, R. J., (2007). One molecule, many derivatives: A never-ending interaction of melatonin with reactive oxygen and nitrogen species? Journal of Pineal Research, 42(1), 28–42.
112. Tang, K. F., Bondad-Reantaso, M. G., Arthur, J. R., MacKinnon, B., Hao, B., Alday-Sanz, V., & Dong, X., (2020). Shrimp acute hepatopancreatic necrosis disease strategy manual. FAO Fisheries and Aquaculture Circular, 1190, 1–65.
113. Terebiznik, M. R., & Pilosof, A. M. R., (1999). Biomass estimation in solid-state fermentation by modeling dry matter weight loss. Biotechnology Techniques, 13(3), 215–219.
114. Thakur, S. A., Nemade, S. N., & Sharanappa, A., (2015). Solid-state fermentation of overheated soybean meal (waste) for production of protease using Aspergillus oryzae. International Journal of Innovative Research in Science, Engineering and Technology, 4, 18456–18461.
115. Thomas, L., Larroche, C., & Pandey, A., (2013). Current developments in solid-state fermentation. Biochemical Engineering Journal, 81, 146–161.
116. Thompson, A. W., Foster, R. A., Krupke, A., Carter, B. J., Musat, N., Vaulot, D., & Zehr, J. P., (2012). Unicellular cyanobacterium symbiotic with a single-celled eukaryotic alga. Science, 337(6101), 1546–1550.
117. Tikhonenkov, D. V., Hehenberger, E., Esaulov, A. S., Belyakova, O. I., Mazei, Y. A., Mylnikov, A. P., & Keeling, P. J., (2020). Insights into the origin of metazoan multicellularity from predatory unicellular relatives of animals. BMC Biology, 18(1), 1–24.
118. Tosti, V., Bertozzi, B., & Fontana, L., (2018). Health benefits of the Mediterranean diet: Metabolic and molecular mechanisms. The Journals of Gerontology: Series A, 73(3), 318–326.
119. Trevizan, L. C., Santos, Jr. D., Samad, R. E., Vieira, Jr. N. D., Nunes, L. C., Rufini, I. A., & Krug, F. J., (2009). Evaluation of laser-induced breakdown spectroscopy for the determination of micronutrients in plant materials. Spectrochimica Acta Part B: Atomic Spectroscopy, 64(5), 369–377.

120. Tudzynski, B., Sharon, A., & Osiewacz, H., (2002). The Mycota X: Industrial Applications (Vol. 10, pp. 78–98). Springer, New York.
121. Vogel, R. F., Hammes, W. P., Habermeyer, M., Engel, K. H., Knorr, D., & Eisenbrand, G., (2011). Microbial food cultures-opinion of the Senate Commission on Food Safety (SKLM) of the German Research Foundation (DFG). Molecular Nutrition and Food Research, 55(4), 654–662.
122. Voidarou, C., Antoniadou, M., Rozos, G., Tzora, A., Skoufos, I., Varzakas, T., & Bezirtzoglou, E., (2021). Fermentative foods: Microbiology, biochemistry, potential human health benefits and public health issues. Foods, 10(1), 69.
123. Vyas, A., Freitas, A. V., Ralston, Z. A., & Tang, Z., (2021). Fission yeast Schizosaccharomyces pombe: A unicellular "micromammal" model organism. Current Protocols, 1(6), e151.
124. Waksman, S. A., & Woodruff, H. B., (1941). Actinomyces antibioticus: A new soil organism antagonistic to pathogenic and non-pathogenic bacteria. Journal of Bacteriology, 42(2), 231–249.
125. Wang, Q. L., Liu, Y. D., Shen, Y. W., Jin, C. Y., Lu, J. S., Zhu, J. M., & Li, S. H., (1991). Studies on mixed mass cultivation of Anabaena spp. (nitrogen-fixing blue-green algae, cyanobacteria) on a large scale. Bioresource Technology, 38(2–3), 221–228.
126. Wang, Z., Zhuge, J., Fang, H., & Prior, B. A., (2001). Glycerol production by microbial fermentation: A review. Biotechnology Advances, 19(3), 201–223.
127. Xiao, J. Z., Takahashi, S., Nishimoto, M., Odamaki, T., Yaeshima, T., Iwatsuki, K., & Kitaoka, M., (2010). Distribution of in vitro fermentation ability of lacto-N-biose I, a major building block of human milk oligosaccharides, in bifidobacterial strains. Applied and Environmental Microbiology, 76(1), 54–59.
128. Xie, J. Z., Chang, H. L., & Kilbane, II, J. J., (1996). Removal and recovery of metal ions from wastewater using biosorbents and chemically modified biosorbents. Bioresource Technology, 57(2), 127–136.
129. Zakaria, Z., Chong, S. F., Zahari, A. R., Fauzi, N. A., & Mohd. Shayuti, S. A., (2014). Growth Kinetic of Fresh and Freeze-Dried Pleurotus Sajor-Caju (Oyster Mushroom) Mycelium for Preservation Study, 594, 196–202.

130. Zeikus, J. G., Jain, M. K., & Elankovan, P., (1999). Biotechnology of succinic acid production and markets for derived industrial products. Applied Microbiology and Biotechnology, 51(5), 545–552.
131. Znad, H., Markoš, J., & Baleš, V., (2004). Production of gluconic acid from glucose by Aspergillus niger: Growth and non-growth conditions. Process Biochemistry, 39(11), 1341–1345.

Chapter 3

Fermentation Miniaturization

CONTENTS

3.1. INTRODUCTION

In the 20th century, the huge gap in the scale of cultivation sustained to prevail amongst the colonies of agar and the Erlenmeyer flask cultures. Most of the laboratories had a slight motivation to decrease volumes of culture well below 50 milliliters, as the quantity of house strains and the variants was instead restricted (Inloes et al., 1983). Throughout the 1990s and 1980s, various novel developments urged the necessity for usage of very high quality and small-scale systems of growth. First of all, various fermentative bioprocedures became completely settled, and it generally became increasingly hard to find considerably improved variants. Practically, this intended screening involving larger libraries of variant, and believing that the enhancement in efficiency of the finest variants would be generally less than 5.1 or 9.99% matched to the strain of parent, needed the procedure of evaluation with low inconsistency. This established the necessity for high-quality systems of cultivation. Similar inclinations were thought for medium improvement projects, where progressively numerous parameters were being verified. Secondly, analytical methods comprising chromatographic techniques like liquid chromatography-mass spectrometry (LC-MS) and gas chromatography, also cell-centered inhibition assays and receptor binding assays, were being enhanced to a degree that these techniques could be carried out with pg (pictogram) or ng (nanogram) amounts of the cell products, which is well-matched with sub ml culture volumes in most of the cases. The more delicate bioassays such as mode of action-centered assays in order to identify antibiotic or anticancerous actions of the microbial metabolites, permitted screening of novel antibiotics and drugs, which are frequently inferior metabolites from fungi or actinomycetes, to be carried out with fewer extracts. Thirdly, projects of screening with very costly medium constituents, like ^{13}C-labeled substrates as utilized in the studies of flux-examination, were fiscally viable when fermented in the small volumes of culture. Some other areas of application comprised the monitoring for novel biocatalysts and the enhancement of particular enzymes through screening large variant libraries (Adrio and Demain, 2014).

Early ad hoc struggles on contraction were carried out in-house in the 1980s by the industries, but methodical information in this field began to gather only after various universities, comprising those in, Zurich, Aachen, and London, began devoted projects of research in the 1990s. Much of the exploration was engrossed on 96- and 24-well MTPs, which were the logical selection since they were consistent and being extensively made practical in some other areas of research. Moreover, the broad range of well-suited types of equipment such as pipetting stations and microplate reader was accessible, which saved time and money. The comparatively large section of this chapter devoted to MTPs reveals their current significance in academia and industry.

3.2. FUNDAMENTALS OF MINIATURIZATION OF CULTURE

The information that the normal microbial cell contains is 1,000-fold smaller as compared to the culture vessel of ten-millimeter diameter. It was implied by microbiologists into considering that the impacts of contraction on the distinct cell are guaranteed to be insignificant (Sender et al., 2016). This perception might be correct for the growth vessels having diameters beyond 10 millimeters, though not correct for the smaller vessels, primarily as an outcome of the surface tension impacts and diverse performance of the gas bubbles (Figure 3.1).

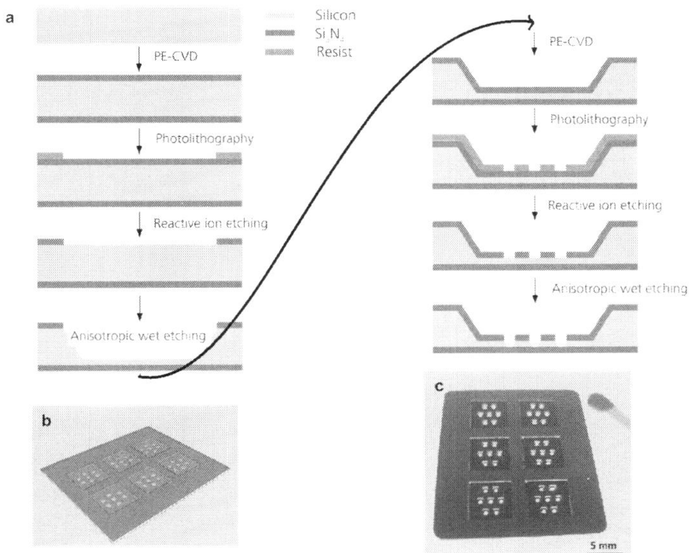

Figure 3.1. The reduced culture compartment of cell. (a) The work pattern of the production. (b) The layout of the microcavity chip. This chip comprises six × seven reduced cell culture compartments. (c) Picture of the MCC [scale bar five millimeter].

Source: https://www.researchgate.net/figure/The-miniaturized-cell-culture-chamber-a-Work-flow-of-the-fabrication-b-Design-of_fig1_51590508.

3.2.1. Effects of the Surface Tension

The surface tension hinders the movement and flow of the medium of culture because of stirring, bubbling, or gravitational forces ascending from the orbital shaking. The significance of the size of the culture becomes obvious when associating the motion of the fluid within vessels with several diameters throughout orbital shaking (Figure 3.2). It became obvious from the photographic studies that impact of the surface tension quickly upsurges at the diameters of vessel below 8-millimeter: at the gravitational force of 2.50, no substantial culture movement of fluid can be developed in the culture vessels of 3 or 4 millimeters. The moderating, even though also obfuscating, the factor is decreasing effect that the culture broth constituents apply on surface tension (Morgan and Shanahan, 1997). The lower surface tension considerably improves the state with the sizes of vessel about 6 millimeters (Figures 3.2 and 3.4), however not with 3 or 4 millimeters. The non-existence of the movement of fluid as an outcome of surface tension impacts outcomes in low rates of transfer of oxygen and other issues.

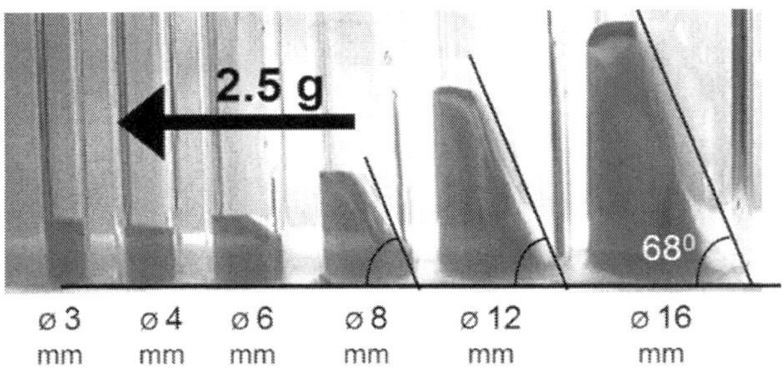

Figure 3.2. High-speed picture displaying the capability of the gravitational force of 2.50 (produced by orbital shaking at 300 revolutions per minute and the shaking diameter of 50 millimeters) to tempt the movement of rotating of the aqueous solution of bromocresol blue in polyacrylate vessels having numerous diameters.

Source: https://onlinelibrary.wiley.com/doi/pdf/10.1128/9781555816827.ch8.

3.2.2. Effects of the Size of Gas Bubble

The performance of the gas bubbles becomes progressively fickle at the smaller sizes of a bubble. Small bubbles incline to arise more gradually and are effortlessly diverged from the precisely vertical movement as compared to the situation with large bubble sizes. The large range of bubble arrangements in identical tiny growth vessels might trigger large dissimilarities in hydrodynamics and transfer rates of oxygen and ultimately to poor reliability of the performance of a process.

3.2.3. Techniques for Determining Transfer Rates of Oxygen

When evaluating the applicability of the new miniature system of cultivation—particularly if someone desires to copycat the physicochemical condition in larger culture volumes—it is desirable to decide transfer rates of oxygen from the gas phase to the liquid phase. Several techniques are accessible to find out transfer rates of oxygen in cultures volumes smaller than one milliliter. The customary and finest-established technique is the Co-catalyzed oxidation of sulfite (Phillips et al., 2011). Here, in this technique, the headspace of a vessel is slowly elated with the oxygen-comprising gas. The medium in a vessel comprises sulfite and Co as the catalyst. All of the oxygen disseminating into a medium is ingested in sulfite oxidation to sulfate so that the concentration of oxygen in the liquid-phase stays close to 0. The transfer rate of oxygen can usually be computed from the change in the concentration of oxygen in the outlet and inlet of gas, or with the help of the pH drop linked with sulfite oxidation (Phillips et al., 2011). The drawback of this technique is the high concentration of salt applied, which decreases the maximal oxygen solubility to a level far below than that in low-salt media, which usually comprises 0.11 to 0.21 M salts. Thus, the maximal transfer rate of oxygen is methodically underrated, and the 21 to 31% factor of correction is essential when interpreting the outcomes to microbial cultures. One more disadvantage of this technique is the high surface tension than the microbial culture, causing an underrate of possibly attainable transfer rates of oxygen in the cultures of small-scale having diameters of generally less than 8 millimeters.

Enzymatic techniques (utilizing oxygen-ingesting enzymes like oxygenases) can be utilized to determine transfer rates of oxygen in the small vessels. The oxidase of glucose in amalgamation with catechol 2,3-dioxygenase and diammonium 2,2-azino-di-(3ethylbenzthiazoline sulfonate) are appropriate, as they give growth to the colored products that are simple to measure spectrophotometrically (Duetz et al., 2000). The technique of C23O is favored since the colored compound is created in the single-stage reaction in a 1:1 ratio to ingest the amount of oxygen. The method of glucose oxidase (GO) displays a less pure stoichiometry and thus needs calibration with other technique, however, has a benefit that all its reagents are commercially available (Duetz et al., 2000).

The third probability is to obtain transfer rates of oxygen from oxygen restricted stage of growth bends of the aerobic strains like Bacillus subtilis or Pseudomonas putida (Druzhinina et al., 2006; Dobson and O'shea, 2008). These later techniques give a more accurate comparison with the real-life circumstances but are more time-wasting and arduous since it is essential to sample the culture at steady intervals of time for 11 to 21 hours.

3.3. CULTIVATION IN MICROTITER PLATES

The orbital shaking is presently the main technique made practical to attain suitable transfer rates of oxygen in microtiter plates. It is famous since the force motivating the upsurge of the collaboration of gas and liquid, as concluded by centrifugal force is identically made practical to all of the wells in all microtiter plates fixed on the platform of quivering, and thus removes the possible basis of well-to-well discrepancy. On the other hand, aeration by magnetic stirring and air sparging nearly unavoidably gives upsurge to well-to-well discrepancies, particularly at the lower volumes of culture (Figure 3.3).

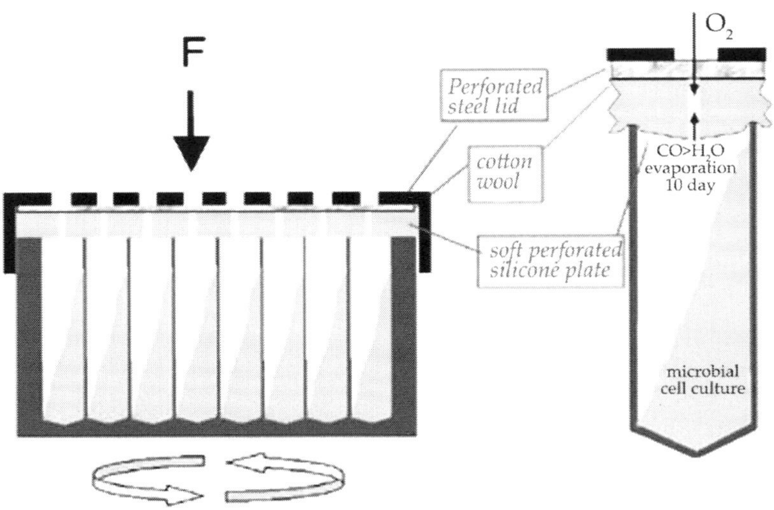

Figure 3.3. Cultivation of cells in MTPs.

Source: https://www.genengnews.com/magazine/51/microbial-and-cell-growth-in-microtiter-plates/.

Throughout shaking of orbital of the microbial cultures in microtiter plates, 2 major factors conclude the reproducibility and quality of the culture: (a) The closure system of well and (b) an amalgamation of the orbital shaking circumstances and volume of culture, which conclude the transfer rates of oxygen and the extent of mixing (Duetz and Witholt, 2001).

3.3.1. Well-Closure Systems

This system of well-closure for the specific wells must satisfy 3 functions: (a) avert (cross)-pollution, (b) allow the swapping of headspace air, and (c) restrict vaporization. Identical physical circumstances in all of the wells are an additional necessity: wells in the corners must have accurately the analogous features as the wells located in the middle of the microtiter plate, and consistency is particularly significant in quantifiable work.

3.3.1.1. Sealing Tape

The gas-porous filters of membrane are accessible from various dealers and are appropriate when utilized in amalgamation with moistened development cabinets as the permeability of water vapor is quite high for all of the sealing tapes given that the fluid of culture does not come in junction with the sealing tape throughout orbital shaking (Dijk et al., 2007). This sealing tape will stay arid when the deep-well microtiter plates are utilized in amalgamation with comparatively small volumes of culture and slight shaking circumstances and when none of the concentration acts. When the sealing tapes are utilized in amalgamation with the high shaking speeds or shallow-well microtiter plates are utilized, though fluid of culture might relax on the foil, lead to numerous hostile effects on the culture quality. First, components of glue might dissolve partly, and if the droplet drops back into culture, it might impact the development or consequent investigative processes. Secondly, the ratio of the area of membrane concealed by the aerosol will vary from one well to the other, backing to the well discrepancy, with respect to gas concentrations and evaporation losses in the headspace. Thirdly, reliant on the sealing tape's quality, the glue might not stick when moistened, chancing cross-pollution and airborne pollution. Besides aerosols, the sealing tapes can generally become wet as an outcome of water condensation. The condensation eagerly happens when this membrane is quite colder as compared to culture (Ivanov et al., 2019). This might happen when the table of shakers is hotter as compared to the air in the cabinet of shakers, which frequently takes place when the motor of shakers heats the table of shakers.

This can be the outcome of the production of heat by the fast-developing cells. In brief, the tapes of sealing are frequently well-matched for qualitative effort under comparatively slighter shaking circumstances, however not under severe circumstances.

3.3.1.2. Sandwich Cover

A normally applicable, although intricate, closure system comprises of the layer of silicone having tiny holes above every well in amalgamation with appropriate layers of the bacterial filters, pushed onto microtiter plate with an adequate force to avert cross-pollution. As the only association with surrounding air is through the tiny holes in the layer of silicon over the middle of every wall, the extent of interchange of the headspace air is completely concluded by the dimensions of the hole. Reliant on the volume of culture and the kind of microbial strain, the appropriate hole dimensions can be chosen. For the fast-developing strains, the rule hired from the area of stirred-tank bioreactors has been confirmed to give suitable outcomes for the small cultures. For the culture volumes of nearly 0.5 milliliters per well, the needed headspace rate of exchange of one-milliliter air every minute can generally be accomplished with holes having a 1.50-millimeter diameter and are 6 millimeters long (Duetz et al., 2000). Lower or higher exchanges of the headspace can generally be accomplished by utilizing smaller or larger holes, correspondingly. The precise exchange rates of headspace air conforming to changing size of hole can normally be determined by measuring the water loss from the well after various incubation days in the cabinet of shaker having a defined humidity and temperature.

Such exchange rate of the headspace-air of one to two volumes of culture every minute will usually keep the oxygen concentration in a headspace above 17.99% under all circumstances and will restrict the loss of culture volume via evaporation to the few percent each day, at 50% surrounding humidity in the cabinet of shaking. The evaporation rate can be lowered by enhancing the humidity in the cabinet of shaking, although one must comprehend that humidities over 74.99% might result in the growth of mold within the cabinet of shaking, which can damage the health of laboratory workers who are opening these shaking cabinets and are unprotected to spores, along with growing levels of contamination (Greeves et al., 1939).

The tiny holes in the layer of silicone of the covers might be congested when dense splashing takes place within a well or generally when the volume of culture is high compared to the complete volume of a well and the shaking condition. It is thus suggested to carry out a variety of tests, like the utilization of diverse volumes, dissimilar shaking circumstances, and examine the sandwich covers for obstruction of these holes.

The benefit of these sandwich covers is that rates of shaking and therefore transfer rates of oxygen can generally be higher. The direct contact between an inert silicone layer and the culture fluid does not have any severe results. Cross-pollution will not occur till the sandwich cover is not taken off during the growth phase. Conversely, if intermediary sampling is needed, one must preferably select an appropriate amalgamation of the volume of culture, kind of MTP, and circumstances of shaking so that the fluid of culture does not touch the silicone layer; else cross-pollution might happen when this sandwich cover is kept back onto microtiter plate after sampling (Federico, 2010).

3.3.2. Transfer Rates of Oxygen

An adequate extent of interchange of the headspace air, which can be accomplished by utilizing sandwich covers or sealing tapes, is the precondition for a good supply of oxygen but does not automatically confirm that the supply of oxygen in a headspace arrives at the cells in suspension. In order to accomplish this, an adequate level of gas to liquid transfer must be established (Figure 3.4).

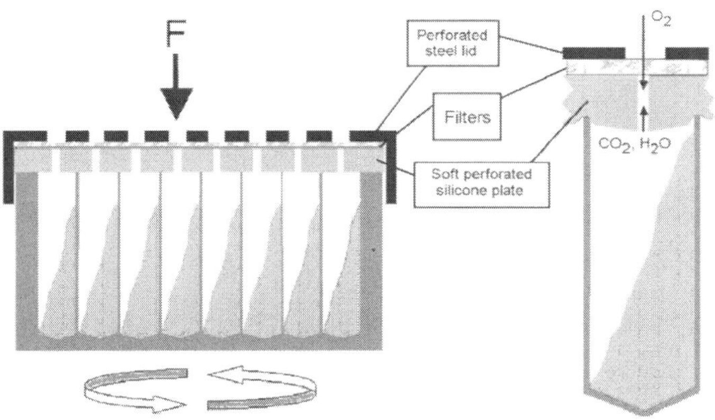

Figure 3.4. Representation of a sandwich cover utilized for (a) the avoidance of cross-pollution of the wells of 96 square deep well microtiter plate throughout dynamic orbital shaking; and (b) the restraint of evaporation.

Source: https://www.researchgate.net/figure/Schematic-view-of-the-sandwich-cover-section-8312-used-for-i-the-prevention-of_fig1_286270336.

The existing agreement in literature is that adequate levels of transfer rates of oxygen (20–51 mmol/liter/hour) can normally be reached when utilizing a comparatively high amplitude of shaking of 24 or 51 millimeters in amalgamation with the shaking frequency of 226 to 300 revolutions per minute (Duetz et al., 2004). For the 24 well plates, good outcomes can be attained with the 51-millimeter amplitude or the 24-millimeter amplitude. For the 96 well microtiter plates, the amplitude of shaking of 51 millimeters gives improved outcomes than 24 mm. This primarily counterintuitive assumption, that is, the smaller the vessel will be, the higher the amplitude of shaker should be to accomplish a reasonable transfer rate of oxygen, becomes reasonable when one comprehends that with the large vessels, the largest section of diameter at which majority of the liquid is whirling essentially curtails from vessel and not from the shaker. The donation of diameter of a vessel to the efficient throw of the majority of the liquid is irrelevant in 96 well microtiter plates. The 2nd clarification for the obvious necessity for the high gravitational force is that this causes an upsurge of the fluid angle with a horizontal plane, which causes the larger percentage upsurge of a gas to liquid area of interface with the 96 well microtiter plates (Glennan, 1996).

3.3.3. Turbulence and Mixing

In the procedure of selecting an appropriate amalgamation of the particular kind of microtiter plate, conditions of shaking, and the culture volume, it is significant to comprehend that an acceptable transfer rate of oxygen does not confirm the good extent of mixing. If methodical mixing is vital, it is often desirable to engage microtiter plates having square wells; these give ascend to the more raging hydrodynamic pattern and permits the higher filling height for the given needed transfer rate of oxygen. Approximately, it can be determined that for the indistinguishable transfer rate of oxygen, microtiter plates with square wells permit 3- to 4 times higher culture volumes for each well than microtiter plates having round wells. The higher turbulence in microtiter plates having square wells can generally be the drawback where shear forces might disturb development negatively or a benefit in a condition where the transfer of mass amongst different phases is restrictive.

3.3.4. Pollutants Percolating From Plastic of Microtiter Plates

An insignificant issue that takes place quite often is the negative impacts on the growth of cells applied by contaminated compounds leaking from polypropylene. The compounds blamable are ultraviolet absorbers and the antioxidants (phosphites and phenols), whiteners (benzoxazoles and coumarins), and the heavy metals such as cobalt, which are frequently utilized as additives in the polypropylene (Duetz et al., 2010). To lessen the poisonous impacts, it is desirable to soak freshly attained polypropylene microtiter plates in a boiling sodium hydroxide (one M) for four hours and hydrochloric acid (1 M, 70°C) for four hours (Druzhinina et al., 2006). As this process is quite arduous, it is desirable to reutilize such treated microtiter plates. Direct extraction of the fermentation broth in microtiter plates having organic solvents might also result in the withdrawal of the plastic components as well as metabolites in a culture extraction. Such kinds of compounds might be perceived by analytical methods, for example, LC-MS, and in some situations might hinder with an examination of the microbial metabolites in extracts (Shen et al., 2021).

Polystyrene microtiter plates are usually non-toxic when utilized off the rack, and are thus frequently favored, particularly in the situation of high-output projects. Though, the selection in polystyrene microtiter plates is restricted; presently no polystyrene square deep plates of well are available commercially, but 96-square narrow wells plates are accessible from numerous dealers.

3.3.5. Applicability of Several Microtiter Plates for Diverse Purposes

In the procedure of selecting an appropriate kind of microtiter plate for a particular use, a variety of factors must be well-thought-out. As discussed above, for all kinds of 24-, 48-, and 96-well microtiter plates, transfer rates of oxygen surpassing 31 mmol liters^{-1} hour^{-1} can be accomplished, so this is not a significant criterion of selection. In most circumstances, the more appropriate principle is the least volume of culture essential for consequent analyzes and examination. Though, if few microliters are needed for examination, it does not automatically recommend that typical 96 narrow plates of well are an accurate selection. It is often desirable to begin executing tests with the variety of diverse kinds of microtiter plates and in order to explicitly associate the difference between autonomous replacements with respect to the appropriate concentration of analyte or some other parameter in order to be well-thought-out for the diverse cultures (Duetz et al., 2010).

As the principle, quite lower nonconformities can usually be accomplished with the comparatively high volumes of culture that are possible in the 24-

well microtiter plates. The fundamental causes for the influence of volume of culture on the nonconformity are badly comprehended, but pipetting inaccuracies and the larger organic discrepancy in small cultures, particularly when the cells do not develop entirely adjourned, have a considerable influence at the smaller volumes. The lesser quantities of biomass utilized for inoculation will unavoidably exhibit the larger discrepancy. The ultimate selection for a particular kind of microtiter plate is frequently a trade-off amongst the output and the standard nonconformity amongst independent replicas (Figure 3.5).

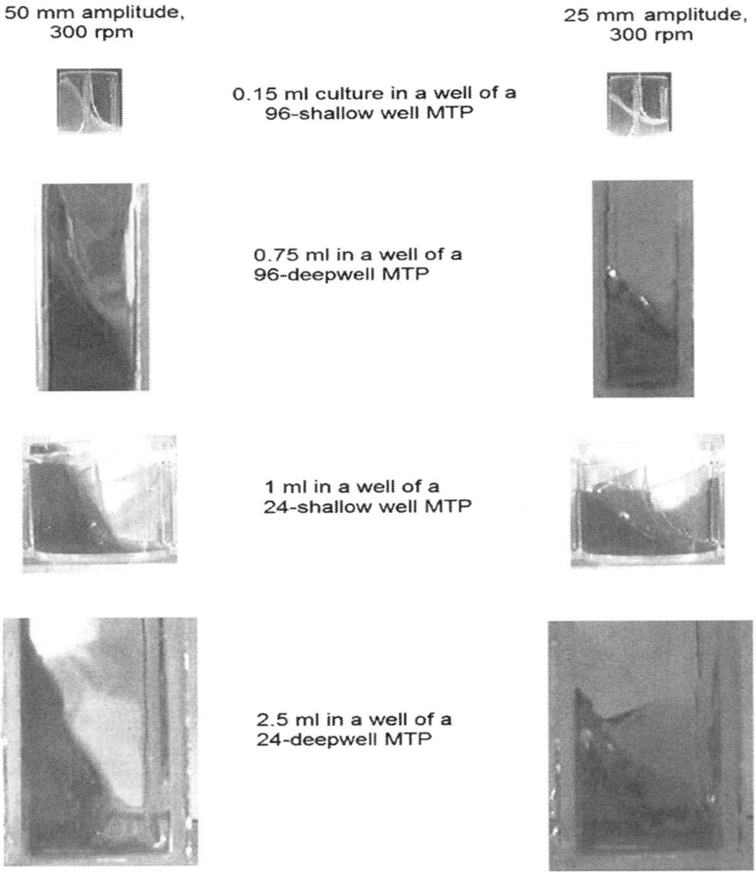

Figure 3.5. The pattern of hydrodynamic flow within wells of several microtiter plates during orbital shaking at almost 300 revolutions per minute, at the shaking amplitude of 25 mm (right) or 50 mm (left).

3.3.6. Management of the Cultures

When presenting mutants libraries, the process of inoculation is vital for consistent outcomes. In several circumstances, the collection of mutants will be primarily accessible as distinct settlements on the plate of agar. The robotic colony selector might be utilized to inject liquid medium distributed in the wells of 96 well microtiter plate. On the other hand, for trivial libraries, one might utilize toothpicks for early inoculation (Barclay, 1987). After a suitable system of well closure has been applied, this initial MTP might be nurtured in the orbital shaker for one to two days. In circumstance of the qualitative selection, for example, to conceal the absence or existence of a specific gene, enzyme, or product, the subsequent suspension of cell can generally be utilized instantly for bioassay or examination. For quantifiable screenings, for example, when locating for high-action variants, it is frequently desirable to utilize the initial MTP in order to inoculate the second microtiter plate, by shifting 1.0 to 5.0 l with the multichannel pipette. The outcome of the initial microtiter plate can usually be kept at 80°C, after adding glycerol. As displayed in Figure 3.6, the cultures in the second microtiter plate will be less or more synchronized. The latter is particularly appropriate if the creation is unsteady or is conspicuously made in the fixed state. The non-synchronized cultures frequently give ascend to numerous incorrect positives and probably incorrect negatives. Practically, harmonizing cultures becomes progressively problematic at smaller culture volumes, most particularly as an outcome of the larger discrepancy in the inoculum's size.

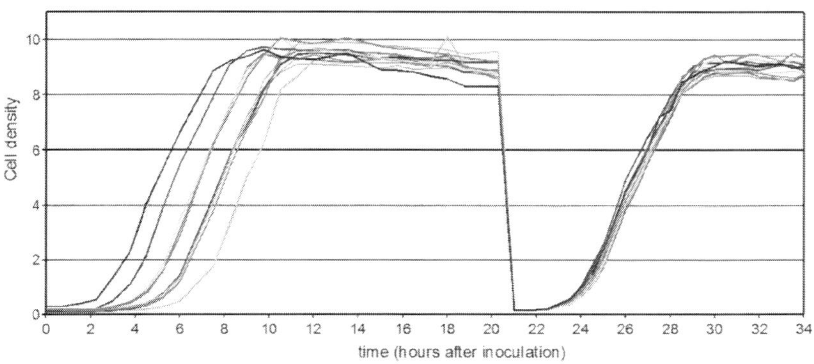

Figure 3.6. Synchronization of 12 isogenic cultures of the E. coli in the wells of the 24 narrow-well microtiter plate throughout orbital shaking (50-millimeter amplitude, 250 revolutions per minute, 30°C).

Note: The initial cultures t → 0 were manually inoculated with the toothpicks from settlements on the LB plate of agar. At t → 20 hours, five-liter of every culture was utilized as the inoculum for 12 new cultures: The 12 cultures now develop in the synchronized pattern, which will in most of the result of the conditions in the smaller standard nonconformity of parameter under research and comparatively ease of recognizing high-action variants.

When beginning with the library in a 96 well microtiter plate format, one can utilize the replicator (96-pin) with stationary pins in order to try the mutant's library after softening the main plate. Otherwise, one can generally press the sterile spring-encumbered replicator onto freezing cultures to obtain the small film of suspension of a cell on pin ends. In any circumstance, one can consequently move the tested cells into the liquid culture directly or on the surface of agar in the rectangular petri dish (Zelle et al., 2010).

3.3.7. Glucose Systems of Feeding

Several large-scale bio procedures are generally systems with fed-batch: when the primarily present glucose is completely ingested, the constant serving of the highly concentrated solution of glucose begins. Variant or medium screenings focused on augmented efficiency of such kinds of systems are preferably carried out in an analogous fed-batch manner. This might be the accurate choice for 96 or 24 cultures utilizing peristaltic pumps or the syringe-centered systems of serving, but such kinds of external systems of dosing are not basically practicable for several cultures. For numerous cultures in microtiter plates, two internal glucose systems of delivery are relevant.

First, tiny silicon elastomer disks comprising crystals of glucose might be added to every well (Hobbs et al., 1989). The added glucose will gradually dispense from these disks into a medium. The 2nd option is the adding of an amalgamation of glucosidases and cyclodextrin or starch to every well. By changing the quantity of glucosidase in a medium, the rate of supply of glucose can be attuned to the level that imitates the large-scale procedure. The 2nd approach was licensed by Rheinwald and Green from MIT in the year 1975 and is currently further expanded upon by Panula-Perälä et al. (2007). The diverse benefits of these silicon elastomer disks are that the compounds excluding glucose can be utilized and that this approach can be made practical for varieties that yield proteases. A logistic and practical benefit of the enzymatic technique is that essential components can generally be added as liquids.

The considerably higher densities of cells that can generally be outreached with the systems make it interesting in order to keep the value of pH in a particular threshold in the lack of the active pH system of control. A comparatively robust buffer is suggested. Also, ammonia (NH_3) is rather not utilized as the source of nitrogen as with ingestion of every molecule of NH_4, 1 proton is discharged and the medium might therefore get acidified quickly. Utilization of the source of energy and acid carbon causes a rise in pH when ingested and might therefore counteract the pH drop because of the ingestion of NH_4. If the sugar is utilized as the source of energy and carbon, applying a high transfer rate of oxygen shaking conditions will usually keep the fermentative creation of acids to the least possible for numerous microbial strains (Bills et al., 2008).

3.3.8. Measurement of O_2, pH, and Biomass Throughout Growth

The tiny cultures size in microtiter plates sets restrictions to the techniques utilized for culture parameters. The utilization of electrode or some other probes is not very useful due to the needed wiring. Due to these reasons, non-invasive techniques like optical sensors are favored.

Often utilized for oxygen and pH calculations are fluorescence-centered sensors composed of tiny, hydrophilic sensing films comprising of the analyte-sensitive indicator and the reference fluorophore that is dropped on the bottom of the microtiter plate.

For biomass quantification within the wells of microtiter plates, optical techniques centered on light scattering can be made practical. Fundamentally three strategies can be eminent. In the first method, the regular microtiter plate bookworm is utilized to trail the optical density, in the absence or existence of the low-orbit shaking. Throughout the computation, orbital shaking is disturbed. The limitation of this method is that transfer rates of oxygen are low, and the unobstructed light transmission becomes nearly 0 at the optical densities usually higher than three, occasioning in the low precision at the high densities of a cell. In the second method, bottoms of the specific wells of the microtiter plate are lightened utilizing the optical fiber bundle (Ortiz et al., 2005). The light intensity reflected by cells developing within the wells is continuously measured utilizing the photodetector. Given that the shaking diameter is not too large, the shaking does not need to be ceased throughout the measurement; therefore, mixing, and supply of O_2 are not disturbed. As one can select diverse wavelengths for light excitation and emission, this method is appropriate for fluorescence computations.

The 3rd method comprises of fixing a flatbed scanning device amongst the MTPs and the shaker (Duetz, 2007). At steady intervals of time, the shaker is ceased, and the bottoms of microtiter plates are skimmed, instantaneously trailed by the automatic resume of the orbital shaker. Consequently, image examination software automatically produces growth curves. The drawback of this 3rd method is the lower supply of oxygen throughout the phase of scanning, which generally can have irregular effects and can trigger stress reactions for definite strains. The benefit, though, is that the consumer is free to select the large diameter of shaking if anticipated. The accessibility of graphics with high resolution of every well permits the customer to study morphological features. The 2nd and the 3rd approach permit the quantification of comparatively high concentrations of biomass equal to 20–50 gram per liter dry weight of the cell.

3.3.9. Particular Features of Growth and the Production Strain

- Fill all of the 96.0 wells of the sterilized 96 deep-well microtiter plates with 0.5 milliliters of a suitable sterile abundant liquid growth medium.
- Inoculate every well with the distinct settlement from the plate of agar spread with a suitably adulterated mutagenized population. Because of the habit of several actinomycete strains developing on agar, the selection of settlements might include taking core models from an agar plate.
- Shield the inoculated microtiter plates with sterilized sandwich covers.
- Incubate the microtiter plate with covers in the orbital shaker at a suitable temperature, normally 26 to 29°C, for 48 hours. Conditions of shaker: 300 revolutions per minute, 50-millimeter shaking amplitude.
- Make the second microtiter plate with a suitably optimized and verified medium of production for a production strain.
- Inoculate the second microtiter plate by removal of 25 l from every well of the initial microtiter plate. For this objective, utilize the 8- or 12-channel pipette having broad-orifice tips. Utilization of robotics might be probable at this step reliant on the setup of a laboratory with respect to sterile practices; medium viscosity and culture morphology are extra factors potentially disturbing utilization of robotics.

- Optional: utilize the initial microtiter plate to make the frozen main plate for feasible rescreening at the later time. Add 100 l every of a 50% solution of glycerol and the main plate culture to every well of the sterilized polystyrene 96 well plate. Utilize an 8- or 12-channel pipette for methodical mixing. Freeze at almost 80°C and store in the cryo box.
- Incubate the production culture microtiter plates in the orbital shaker at a suitable temperature, normally 26 to 29°C, for a suitable length of time for a specific production strain. Conditions of shaker: 300 revolutions per minute, 50-millimeter shaking amplitude.
- Examine inhabitants for titer, action, or other-directed standards:
  - Cell-related natural products: implement cells extractions as essential from filtered or centrifuged mass in order to mine compounds of curiosity. Utilize extracts for HPLC, GC-MS, or cellular-centered examination of the titer or action.
  - Extracellular products: gather supernatant by filtration or centrifugation.
  - Substitutions:
  - Extract complete well utilizing liquid-liquid techniques of extraction with a suitable water-immiscible solvent like ethyl acetate.
  - Smear complete well fillings to suitable solid support like Celite-filled 96 well devices tracked by liquid-to-liquid methods of extraction with a suitable water-unmixable solvent like ethyl acetate.
  - Smear supernatant models to suitable 96 well solid-phase devices of extraction trailed by suitable abstraction and, if required, concentration steps.

3.4. CULTIVATION IN TEST TUBES

Whereas they are not intrinsically parallelized systems, one should not exclude the usefulness of test tubes in order to culture large to small amounts of the microbial strains, particularly if someone desires to manage all of the cultures separately. Test tubes are utilized for several purposes, but here, emphasis is laid on the utilization of the tubes for high output culturing of the microbes in very small volumes. 1,237 actinomycete strains were screened by Ohta et al. for the capability to hydroxylate cyclosporine A enzymatically, therefore imitating human metabolism of the intricate structure (Minas et al., 2000). In this situation, all of the strains were protected from the freeze vegetative mycelial piles into two milliliters of a medium in the test tubes, and the test tubes were usually grown on the rotary shaker at 211 revolutions per minute. At last, almost 89 strains developed and selected in this particular way were discovered to yield the hydroxylated metabolites of concern, and numerous were effectively scaled up to 31-liter jar fermentors. Screening of 250 different actinomycete strains in the cultures of test tube using 10 diverse media of fermentation in order recognize HPLC profiles of the metabolites yielded was reported by Tormo et al. (2003) This was anticipated to recognize the media usually compassionate of a broader chemical multiplicity of the yields (Samorski et al., 2005). Every strain in particular task was inoculated into 10 milliliters of every medium of creation in 26-by-151-millimeter test tubes, and these cultures were generally developed in test tubes at the 35° angle. The outcomes of the strains screening and several media of production were finally compassionate of the 1 strain-numerous compounds strategy to a screening of microbial natural product and also validated the usefulness of these test tubes in very high output microbial screening (Figure 3.7) (Bochner, 2009).

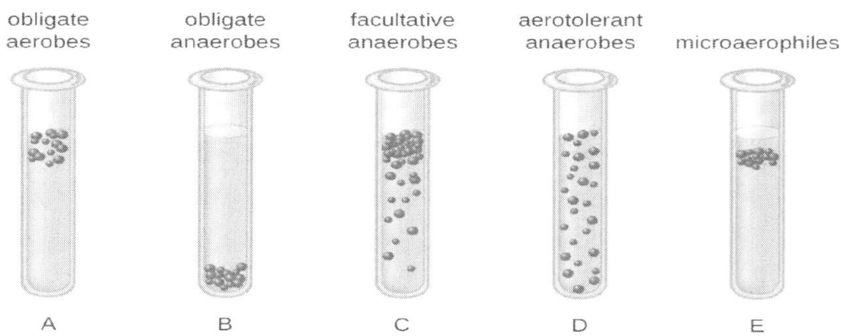

Figure 3.7. Illustration of the distribution of bacterial cells in the thioglycolate tubes.

Source: https://courses.lumenlearning.com/microbiology/chapter/oxygen-requirements-for-microbial-growth/.

3.5. GOVERNED MINI-BIOREACTOR SYSTEMS

The utilization of these systems like microtiter plates has helped scientists in a variety of microbiological uses. The chapter has engrossed on microtiter plates of several well volumes and geometries, in the lack of medium, O_2, or pH control. Though, there prevails a variety of substitute mini-bioreactor systems having prospective for numerous uses (Figure 3.8).

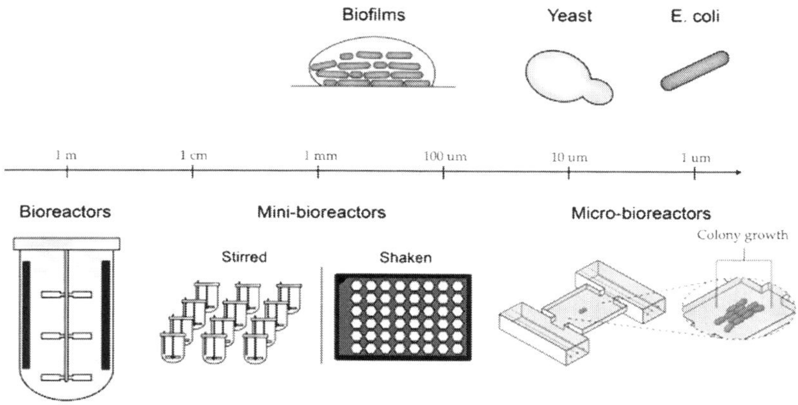

Figure 3.8. Summary of diverse bioreactor technologies accessible, from large to micro and mini-scale, and contrast with a typical size of the biological systems.

Source: https://www.sciencedirect.com/science/article/pii/B978044463663800015X.

From simple to superb, the literature comprises of various instances of the other fermentation and mini-bioreactor systems. Baganz and Betts present a broad assessment of reduced bioreactors and discover the benefits and drawbacks of various existing and evolving technologies (Betts and Baganz, 2006). As described in the assessment, there might be a benefit to choosing each kind of micro- or mini-bioreactor centered on the application. Two particular instances of these systems are defined below.

Puskeiler et al. (2005) developed the block of reaction having 48 mini-reactors with the operating volumes of 8.0–12.0 milliliter each, and self-centering and free-floating magnetic impellers (Ondeyka et al., 2009). The impeller's shape in amalgamation with perplexes triggered the air in order to be drawn into the medium, causing maximal transfer rates of oxygen of almost 301 mmol liter^{-1} hour^{-1}. The robotic system of liquid handling permits the systems to operate as a set of 48 completely governed fed-batch bioreactors.

The tiny bioreactor system defined by Isett et al. (2007) comprises of the altered polystyrene 24 well microtiter plate. Oxygen and pH are constantly monitored utilizing fluorescence-centered sensors. The pH generally can be adjusted automatically to the set level by ammonia or CO_2 through the gas phase. The orbital shaking at 499 to 799 revolutions per minute enables transfer rates of oxygen in the range of 11–16 mmol $liter^{-1}$ $hour^{-1}$. Alternatively, air sparging can generally be utilized to supply additional oxygen to the cultures (Isett et al., 2007). This system was assessed with E. coli, Pichia pastoris, and Saccharomyces cerevisiae strains.

Even though it was not contracted to an extent as the two examples mentioned above, it is also anticipated to mention here the advancement of the shaken-flask parallel system (Anderlei et al., 2004). Utilization of this system permits for the growth of microorganisms in 14 parallel shaking flasks: six typical laboratory flasks in order sample along with eight dedicated flasks for the online measurements of transfer rates of oxygen, respiratory quotient, and the transfer rates of CO_2. This system also permits for screening of several conditions, like medium composition, whereas acquiring information about the organism's metabolism.

REFERENCES

1. Adrio, J. L., & Demain, A. L., (2014). Microbial enzymes: Tools for biotechnological processes. Biomolecules, 4(1), 117–139.
2. An, Z., Harris, G., Zink, D., Giacobbe, R., Lu, P., Sangari, R., Bills, G., Svetnik, V., Gunter, B., Liaw, A., Masurekar, P., Liesch, J., Gould, S., & Strohl, W., (2005). Expression of cosmid-size DNA of slow-growing fungi in Aspergillus nidulans for secondary metabolite screening. Handbook of Industrial Mycology, 12(1), 167–187.
3. Anderlei, T., Zang, W., Papaspyrou, M., & Buchs, J., (2004). Online respiration activity measurement (OTR, CTR, RQ) in shake flasks. Biochem. Eng. J., 17, 187–194.
4. Arain, S., John, G. T., Krause, C., Gerlach, J., Wolfbeis, O. S., & Klimant, I., (2006). Characterization of microtiter plates with integrated optical sensors for oxygen and pH, and their applications to enzyme activity screening, respirometry, and toxicological assays. Sensors Actuators, 113, 639–648.
5. Barclay, H. J., (1987). Models for pest control: Complementary effects of periodic releases of sterile pests and parasitoids. Theoretical Population Biology, 32(1), 76–89.
6. Betts, J. I., & Baganz, F., (2006). Miniature bioreactors, current practices, and future opportunities. Microb. Cell Fact., 5, 21.
7. Bills, G. F., Martín, J., Collado, J., Platas, G., Overy, D., Tormo, J. R., Vicente, F., Verkleij, G., & Crous, P., (2009). Measuring the distribution and diversity of antibiosis and secondary metabolites in the filamentous fungi. Soc. Ind. Microbiol. News, 59, 133–146.
8. Bills, G. F., Platas, G., Fillola, A., Jiménez, M. R., Collado, J., Vicente, F., & Peláez, F., (2008). Enhancement of antibiotic and secondary metabolite detection from filamentous fungi by growth on nutritional arrays. Journal of Applied Microbiology, 104(6), 1644–1658.
9. Bills, G. F., Platas, G., Overy, D. P., Collado, J., Fillola, A., Jiménez, M. R., Martín, J., González Del Val, A., Vicente, F., Tormo, J. R., Peláez, F., Calati, K., Harris, G., Parish, C., Xu, D., & Roemer, T., (2009). Discovery of the parnafungins, antifungal metabolites that inhibit mRNA polyadenylation, from the Fusarium larvarum complex and other Hypocrealean fungi. Mycologia, 101, 449–472.

10. Bills, G. F., Platas, G., Overy, D. P., Collado, J., Fillola, A., Jimenez, M. R., & Roemer, T., (2010). Antifungal metabolites to inhibit mRNA polyadenylation, Mycologia, 100(2), 4–12.
11. Bochner, B. R., (2009). Global phenotypic characterization of bacteria. FEMS Microbiol. Rev., 33,191–205.
12. Bode, H. B., Bethe, B., Höfs, R., & Zeeck, A., (2002). Big effects from small changes, possible ways to explore nature's chemical diversity. Chembiochem., 3, 619–627.
13. De la Cruz, T. E. E., Schulz, B. E., Kubicek, C. P., & Druzhinina, I. S., (2006). Carbon source utilization by the marine Dendryphiella species D. arenaria and D. salina. FEMS Microbiol. Ecol., 58, 343–353.
14. Dobson, L. F., & O'Shea, D. G., (2008). Antagonistic: Effect of divalent cations Ca^{2+} and Mg^{2+} on the morphological development of Streptomyces hygroscopicus var. geldanus. Appl. Microbiol. Biotechnol., 81,119–126.
15. Doig, S. D., Diep, A., & Baganz, F., (2005). Characterization of a novel miniaturized bubble column bioreactor for high throughput cell cultivation. Biochem. Eng. J., 23, 97–105.
16. Druzhinina, I. S., Schmoll, M., Seiboth, B., & Kubicek, C. P., (2006). Global carbon utilization profiles of wild-type, mutant, and transformant strains of Hypocrea jecorina. Appl. Environ. Microbiol., 72, 2126–2133.
17. Duetz, W. A., & Witholt, B., (2001). Effectiveness of orbital shaking for the aeration of suspended bacterial cultures in square-deep-well microtiter plates. Biochemical Engineering Journal, 7(2), 113–115.
18. Duetz, W. A., & Witholt, B., (2004). Oxygen transfer by orbital shaking of square vessels and deep-well microtiter plates of various dimensions. Biochem. Eng. J., 17, 181–185.
19. Duetz, W. A., (2007). Microtiter plates as mini-bioreactors, miniaturization of fermentation methods. Trends Microbiol., 15, 470–475.
20. Duetz, W. A., (2008). An apparatus and a method for investigation of microtiter plates subjected to orbital shaking. Netherlands Patent, 86(1), 881.
21. Duetz, W. A., Chase, M., & Bills, G., (2010). Miniaturization of fermentations. Manual of Industrial Microbiology and Biotechnology, 2, 99–116.

22. Duetz, W. A., Rüedi, L., Hermann, R., O'Connor, K., Büchs, J., & Witholt, B., (2000). Methods for intense aeration, growth, storage, and replication of bacterial strains in microtiter plates. Appl. Environ. Microbiol., 66, 2641–2646.
23. Federico, S., (2010). Outsourcing versus integration at home or abroad and firm heterogeneity. Empirica, 37(1), 47–63.
24. Frisvad, J. C., & Samson, R. A., (2004). Polyphasic taxonomy of Penicillium subgenus Penicillium: A guide to identification of food and air-borne Terverticillate penicillia and their mycotoxins. Stud. Mycol., 49, 1–173.
25. Glennan, S. S., (1996). Mechanisms and the nature of causation. Erkenntnis, 44(1), 49–71.
26. González del Val, Jiang, B., Kahn, J. N., Galuska, S., Giacobbe, R. A., Abruzzo, G. K., Hickey, E. J., Liberator, P. A., Roemer, T., & Singh, S. B., (2009). Isolation, structure, and biological activity of phomafungin, a cyclic lipodepsipeptide from a widespread tropical Phoma sp. Bioorg. Med. Chem., 17, 1361–1369.
27. Green, H., & Rheinwald, J. G., (1975). Method of controllably releasing glucose into a cell culture medium. U.S. Patent, 3(926), 723.
28. Greeves-Carpenter, C. F., Vetter, H., Chamberlain, A. B., Baker, C. T., Sprague, R., Avery, L. T., & MacArthur, R. A., (1939). Refrigeration Engineering, 37, 79.
29. Hermann, R., Lehmann, M., & Büchs, J., (2003). Characterization of gas-liquid mass transfer phenomena in microtiter plates. Biotechnol. Bioeng., 81, 178–186.
30. Hermann, R., Walther, N., Maier, U., & Büchs, J., (2001). Optical method for the determination of the oxygen-transfer capacity of small bioreactors based on sulfite oxidation. Biotechnol. Bioeng., 74, 355–363.
31. Hobbs, G., Frazer, C. M., Gardner, D. C. J., Cullum, J. A., & Oliver, S. G., (1989). Dispersed growth of Streptomyces in liquid culture. Appl. Microbiol. Biotechnol., 31, 272–277.
32. Inloes, D. S., Taylor, D. P., Cohen, S. N., Michaels, A. S., & Robertson, C. R., (1983). Ethanol production by Saccharomyces cerevisiae immobilized in hollow-fiber membrane bioreactors. Applied and Environmental Microbiology, 46(1), 264–278.

33. Isett, K., George, H., Herber, W., & Amanullah, A., (2007). Twenty-four-well plate miniature bioreactor high throughput system, assessment for microbial cultivations. Biotechnol. Bioeng., 98, 1017–1028.
34. Ivanov, P., Kedersha, N., & Anderson, P., (2019). Stress granules and processing bodies in translational control. Cold Spring Harbor Perspectives in Biology, 11(5), a032813.
35. Jeude, M., Dittrich, B., Niederschulte, H., Anderlei, T., Knocke, C., Klee, D., & Buchs, J., (2006). Fed-batch mode in shake flasks by slow-release technique. Biotechnol. Bioeng., 95, 433–445.
36. Junker, B. H., Heese, M., Burgess, B., Masurekar, P., Connors, N., & Seely, A., (2004). Early phase process scale-up challenges for fungal and filamentous bacterial cultures. Appl. Biochem. Biotechnol., 119, 241–277.
37. Kensy, F., Zimmermann, H. F., Knabben, I., Anderlei, T., Trauthwein, H., Dingerdissen, U., & Büchs, J., (2005). Oxygen transfer phenomena in 48-well microtiter plates: Determination by optical monitoring of sulfite oxidation and verification by real-time measurement during microbial growth. Biotechnology and Bioengineering, 89(6), 698–708.
38. Kubicek, C. P., Bissett, J., Druzhinina, I., Kullnig-Gradinger, C., & Szakacs, G., (2003). Genetic and metabolic diversity of Trichoderma, a case study on South-East Asian isolates. Fungal Genet. Biol., 38, 310–319.
39. Minas, W., Bailey, J. E., & Duetz, W., (2000). Streptomycetes in micro-cultures, growth, production of secondary metabolites, and storage and retrieval in the 96-well format. Antonie van Leeuwenhoek, 78, 297–305.
40. Monaghan, R. L., Gagliardi, M. M., & Streicher, S. L., (1999). Culture preservation and inoculum development. Manual of Industrial Microbiology and Biotechnology, 5, 29–48.
41. Morgan, M., & Shanahan, J., (1997). Two decades of cultivation research: An appraisal and meta-analysis. Annals of the International Communication Association, 20(1), 1–45.
42. O'Cleirigh, C., Casey, J. T., Walsh, P. K., & O'Shea, D. G., (2005). Morphological engineering of Streptomyces hygroscopicus var. geldanus, regulation of pellet morphology through manipulation of broth viscosity. Appl. Microbiol. Biotechnol., 68, 305–310.

43. Ohta, K., Agematu, H., Yamada, T., Kaneko, K., & Tsuchida, T., (2005). Production of human metabolites of cyclosporin A, AM1, AM4N and AM9, by microbial conversion. J. Biosci. Bioeng., 99, 390–395.
44. Ondeyka, J. G., Zink, D., Basilio, A., Vicente, F., Bills, G., Diez, M. T., & Singh, S. B., (2007). Coniothyrione, a chlorocyclopentandienylbenzopyrone as a bacterial protein synthesis inhibitor discovered by antisense technology. Journal of Natural Products, 70(4), 668–670.
45. Ondeyka, J., Harris, G., Zink, D., Basilio, A., Vicente, F., Bills, G., Platas, G., Collado, J., González, A., M. de la Cruz, Martín, J., Kahn, J. N., Galuska, S., Giacobbe, R., Abruzzo, G., Hickey, E., Liberator, P., Jiang, B., Xu, D., Roemer, T., & Singh, S. B., (2009). Isolation, structure elucidation and biological activity of virgineone from Lachnum virgineum using the genome-wide Candida albicans fitness test. J. Nat. Prod., 72, 136–141.
46. Ortiz-Ochoa, K., Doig, S. D., Ward, J. M., & Baganz, F., (2005). A novel method for the measurement of oxygen mass transfer rates in small-scale vessels. Biochem. Eng. J., 25, 63–68.
47. Panula-Perälä, J., Siurkus, J., Vasala, A., Wilmanowski, R., Casteleijn, M. G., & P. Neubauer., (2008). Enzyme controlled glucose auto-delivery for high cell density cultivations in microplates and shake flasks. Microb. Cell Fact., 7, 31–42.
48. Phillips, J. W., Goetz, M. A., Smith, S. K., Zink, D. L., Polishook, J., Onishi, R., & Singh, S. B., (2011). Discovery of kibdelomycin, a potent new class of bacterial type II topoisomerase inhibitor by chemical-genetic profiling in Staphylococcus aureus. Chemistry and Biology, 18(8), 955–965.
49. Puskeiler, R., Kusterer, A., John, G. T., & Weuster-Botz, D., (2005). Miniature bioreactors for automated high throughput bioprocess design (HTBD), reproducibility of parallel fed-batch cultivations with Escherichia coli. Biotechnol. Appl. Biochem., 42, 227–235.
50. Samorski, M., Muller-Newen, G., & Buchs, J., (2005). Quasi-continuous combined scattered light and fluorescence measurements, a novel measurement technique for shaken microtiter plates. Biotechnol. Bioeng., 92, 61–68.
51. Sender, R., Fuchs, S., & Milo, R., (2016). Revised estimates for the number of human and bacteria cells in the body. PLoS Biology, 14(8), 1002533.

52. Shen, Y., Borowski, J. E., Hardy, M. A., Sarpong, R., Doyle, A. G., & Cernak, T., (2021). Automation and computer-assisted planning for chemical synthesis. Nature Reviews Methods Primers, 1(1), 1–23.
53. Smedsgaard, J., (1997). Micro-scale extraction procedure for standardized screening of fungal metabolite production in cultures. J. Chromatogr. A., 760, 264–270.
54. Talbot, N. J., Vincent, P., & Wildman, H. G., (1996). The influence of genotype and environment on the physiological and metabolic diversity of Fusarium compactum. Fungal Genet. Biol., 20, 254–267.
55. Tormo, J. R., Garcia, J. B., M. DeAntonio, Feliz, J., Mira, A., Diez, M. T., Hernandez, P., & Pelaez, F., (2003). A method for the selection of production media for actinomycete strains based on their metabolite HPLC profiles. J. Ind. Microbiol. Biotechnol., 30, 582–588.
56. Van Dijk, R., Slump, E., Meens, M., Joosten, R., & Emalfarb, M., (2007). A dedicated vector for efficient library construction and high throughput screening in the hyphal fungus Chrysosporium lucknowense. Indust. Biotechnol., 3, 48–57.
57. Van Wezel, G. P., Krabben, P., Traag, B. A., Keijser, B. J. F., Kerste, R., Vijgenboom, E., Heijnen, J. J., & Kraal, B., (2006). Unlocking Streptomyces spp. for use as sustainable industrial production platforms by morphological engineering. Appl. Environ. Microbiol., 72, 5283–5288.
58. Vicente, Martín, J., González, A., J. Bur-Zimmermann, Tormo, J. R., & Peláez, F., (2008). Enhancement of antibiotic and secondary metabolite detection from filamentous fungi by growth on nutritional arrays. J. Appl. Microbiol., 104, 1644–1658.
59. Vinci, V. A., & Byng, G. R. A. H. A. M., (1999). Strain improvement by non-recombinant methods. Manual of Industrial Microbiology and Biotechnology, 2, 103–113.
60. Visser, H., Joosten, V., Punt, P. J., Gusakov, A. V., Olson, P. T., Joosten, R., & Wery, J., (2011). Development of a mature fungal technology and production platform for industrial enzymes based on a Myceliophthora thermophila isolate, previously known as Chrysosporium lucknowense C1. Industrial Biotechnology, 7(3), 214–223.
61. Whitaker, A., (1992). Actinomycetes in submerged culture. Appl. Biochem. Biotechnol., 32, 23–35.

62. Wierschke, J. D., Newman, A. S., Brady, C. A., Barnes, N. M., & Guzzo, P. R., (2013). Selected AMRI publications and patents from 2005–2014. Development, 17, 915–920.

63. Yang, W., Hartwieg, E. A., Fang, A., & Demain, A. L., (2003). Effects of carboxymethylcellulose and carboxypolymethylene on morphology of Aspergillus fumigatus NRRL 2346 and fumagillin production. Curr. Microbiol., 46, 24–27.

64. Zelle, R. M., De Hulster, E., Kloezen, W., Pronk, J. T., & Van Maris, A. J., (2010). Key process conditions for production of C4 dicarboxylic acids in bioreactor batch cultures of an engineered Saccharomyces cerevisiae strain. Applied and Environmental Microbiology, 76(3), 744–750.

65. Zimmermann, H. F., John, G. T., Trauthwein, H., Dingerdissen, U., & Huthmacher, K., (2003). Rapid evaluation of oxygen and water permeation through microplate sealing tapes. Biotechnol. Prog., 19, 1061–1063.

Chapter 4

Aerobic and Anaerobic Solid-Phase Fermentation

CONTENTS

4.1. INTRODUCTION

The science of SPF (solid-phase fermentation) outlines the similar prehistoric background that forms the basis of the biotechnology field and also the recent techniques used for microorganisms' growth. This method for microorganism culturing had been utilized to produce pharmaceuticals, diverse foods, chemicals, beverages, high-value biochemical, and commodity enzymes. The discovery of novel natural products works as a beneficial research tool. A former research by researchers covers little scale of SPF, also further assessments are done on SPF (Figure 4.1) (Aidoo et al., 1982; Barrios et al., 1996).

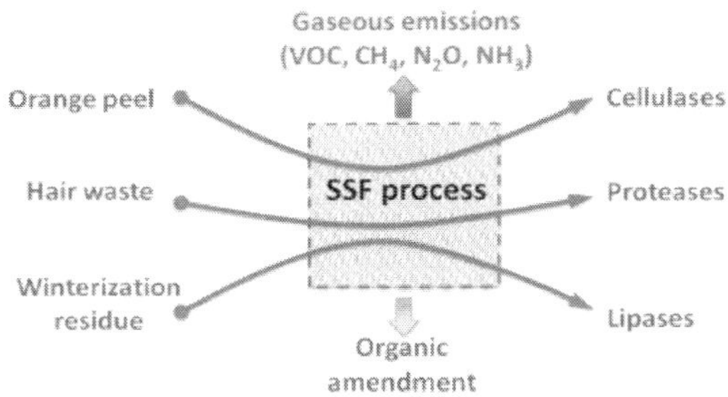

Figure 4.1. The simple solid-state fermentation procedure.

Source: https://www.sciencedirect.com/science/article/abs/pii/S0960852414017684.

Background of fermentation dates back to the early history and provided rise to the initial, possibly accidental, techniques utilized for food preservation and avoid them from indulging and causing sickness (Beuchat, 1987). SPF is as ancient as leavened bread produced in prehistoric Egypt (2600 BC), also beer-produced in Babylonia (6000 BC), centuries old fermentation of soy in cultures of South Asia. Also, the diversity of fermentation of customary food spread globally along with basic agriculture.

The process of koji, a type of SPF and a groundbreaking development, was initiated in China, and in the 7th century, it was taken to Japan as fermentation on steamed rice through Aspergillus oryzae. It was utilized as an "appetizer," however it is the initial phase in the next fermentation that makes sake, miso, tempeh, soy sauce, also other normal oriental diets.

Appropriately the word Kabi-Tachi is a Japanese word whose abbreviation is "koji" and "bloom of the mold" is its meaning. The spread and progress in the solid phase fermentation history were inspirational events (Figure 4.2).

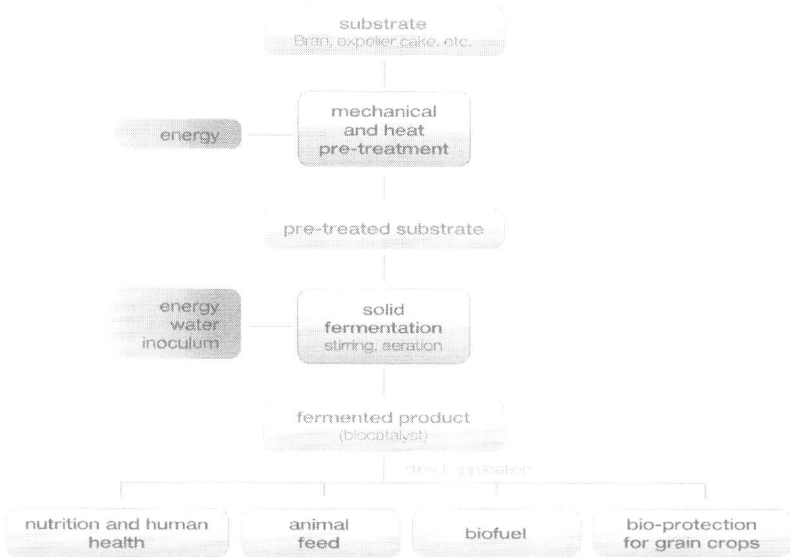

Figure 4.2. Flow cart of solid-state fermentation procedure.

Source: https://en.wikipedia.org/wiki/Solid-state_fermentation.

4.2. CHARACTERISTICS OF SPF

Inclusively it is already defined, SPF in rather anaerobic or aerobic situations, is a procedure of microorganisms culturing on hard ingredients in the presence of no or little unrestricted H_2O. Hard ingredients on which they flourish, or might enter, normally have water (and frequently solvable nutrients) or they can engross it. Usually, these substrates are legumes, numerous lignocellulosic, and cereal grains. It could be supposed that SPF involves both solid-state fermentation (SSF) and solid substrate working as normal ingredient as it is a source of carbon/energy in the presence of no or little unrestricted H_2O. However, in the wider description, SPF, comprises the probable application of a degradable, slow, or poorly degradable substrate in an appropriate medium of hard support (Figure 4.3).

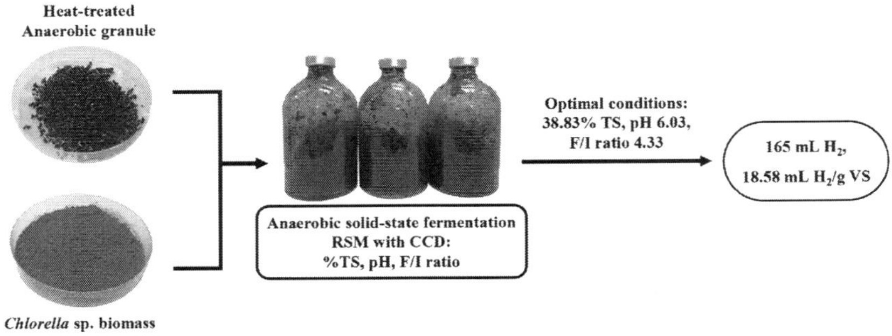

Figure 4.3. Anaerobic solid-state fermentation of biohydrogen.

Source: https://www.sciencedirect.com/science/article/abs/pii/S0360319917302513.

Additionally, inert backings (Bigelis et al., 2006; Ooijkaas et al., 2000) could play a significant role in SPF and forms its classification and descriptions. SPF could be subcategorized as per the nature of microorganisms used for the fermentation: mixed culture fermentation, natural indigenous culture, and fermentation of uncontaminated monoculture. Usually, pure described cultures are utilized in most industrial SPF as this method is related to reproducibility and better-quality control. However, certain fermentations with food applications and remains of agriculture work combined cultures with advantages. Generally, meanings of all SPF discriminate its several types from the immersed fermentation, called LPF (liquid phase fermentation) or liquid fermentation (LF) which is carried out in dilute solutions or slurries, in comparison to SPF, where no or slight liquid is existing (Bigelis et al., 2006). In industrial microbiology, LPF is of main importance as it is utilized to create several and diverse commercial products. An additional method, by Bigelis et al. (2006) is called MPF (mixed-phase fermentation), ties SPF and LPF, and employs permeable inert solid backings in a plentiful liquid medium to assist the novel natural products discovery.

4.3. ADVANTAGES AND DISADVANTAGES OF SPF

4.3.1. Advantages

SPF might offer benefits, particularly savings cost, above immersed fermentation (Balakrishnan and Pandey, 1996). In different circumstances, SPF had been presented to be a normal technology. The features are, it is:

(i) less vulnerable to impurity through unwanted organisms because of little activity of water (a_w); (ii) more productive; (iii) autonomous of mixing and agitation; also (iv) cost-efficient as: (a) bareness of substrate (also autoclaving and pasteurization) might be unnecessary; (b) culture media might be plain, comprising of remains of agro-industrial; also (c) airing requirements might decrease (Corry, 1987). Moreover, SPF could utilize fermentors of decreased volume and size with associated decreased water requirements and effluent generation facilities of compact fermentation, and easing culture situations similar to a usual environment needing decreased processing also retrieval through solvents and reagents (Hendry and Wood, 1982; Roussos et al., 1997). Processing of fermented remains is normally very simple, merely needs speedy desiccating of the less humid product to generate fertilizer, feed of animal, or additional dehydrated profitable products (Sato and Sudo, 1999). Lastly, the appropriate creation of conidiophores through the koji technique delivers eagerly accessible initiator cultures. These microorganisms are more steady, vigorous, later cryostorage, and resilient to dryness than microorganisms got through LPF.

SPF is particularly beneficial in applications of fungus, also it is additionally normal with this group of microorganisms. Less moisture content of several hard media utilized for SPF makes it particularly appropriate for molds culturing and the development of filamentous fungi regardless of bacteria, might be restricted through low a_w (Sato and Sudo, 1999). SPF, which is generally a stationary procedure, often evades the mechanical forces feature and shear stress of LPF and mycelia is disturbed. As a consequence, SPF might promote situations for extreme creation. Furthermore, fungal SPF had a record of better production of secondary metabolites (Acuna et al., 1994; Aguilar et al., 2004).

4.3.2. Disadvantages

Drawbacks of SPF might offset the advantages and may also force strategies for alternate fermentation. Absence of scalability in this technique of fermentation often restricts its application. An additional major disadvantage is the less thermal conductivity of the hard substrate, which creates difficulty to regulate the temperature. A rise in temperature, resultant from the heat of metabolics, could form local warm spots rather than forced airing and the thermal controlling devices applications (Saucedo et al., 1990). Further drawbacks are associated with procedure control and agitation: complications in the blending of the substrate bed might lead to irregular water distribution and nonhomogeneity, heat, pH regulating agents, biomass, and nutrients

(Sato and Sudo, 1999). Due to the poor mixing, diffusion of the product might be restricted and interfere through the preferred physiological state and guideline for creating organism pathway. Several additional drawbacks might involve: (i) trouble in measuring and checking directly bacterial cell development with fermentation limits; (ii) incapability to regulate crucial fermentation limits; (iii) wasteful necessities for the substrates and solid ingredients pretreatment; (iv) contamination of the last product raising the costs of retrieval of product; (v) substantial humidity loss throughout long-enduring fermentations; (vi) requirement for the big inocula; also (vii) restrictions because of the crucial function of a_w and water (Sato and Sudo, 1999).

4.4. PROCESS CONTROL

Main SPF procedure variables are associated with the microorganism and its cultivation, the selection of substrate and the pretreatment, also environmental causes which decide that how frequently in a huge scale bioreactor, the substrate and the microorganism interrelate (Krishna, 2005; Mitchell et al., 2006). Effective procedure control relies on model-founded approaches expressive of thermal transference and mass procedures (Sato and Sudo, 1999).

4.4.1. Making of the Organism and Inoculation

The procedures of inoculation take the bacterium, also the cellular state (vegetative or spores cells), the inoculum density, choice of medium, and method of application, the state and nature of the substrate (a_w, mixability, sterility, pH, temperature,) into account. Likened to somatic mycelia, microorganisms are regularly favored as they provide suitability; sustainability, and storing benefits; also, easiness of manipulation, quantitation, and transference, frequently through inoculation of spray (Larroche, 1996). Multiple development centers are also provided. Regardless of their toughness, spores might show holdup times as they develop out of inactivity, also need greater inocula, and development situations reasonably changed from those for vegetative development.

Contrasting LPF where sterility is frequently needed, pure culture methods might not be necessary for SPF, particularly if procedure bacteria are capable to outcompete the polluting bacteria. Though hygiene is necessary, a sterile operation might not be essential, also the price might be reduced (Pandey et al., 2000).

4.4.2. Making of Materials

Substrates of SPF and initiating materials might be pretreated prior to inoculation, when required, to assist diffusion of mycelial, and however conditioned to encourage moistening, aeration, mixing nutrients, or gases diffusion. Dehulled rice, a general substrate for the koji procedure, for instance, might be refined to eliminate nearly 41%–70% of the mass, and afterward steamed, increasing its water quantity 10%–15% by mass. Pretreatment of steam is utilized for substrate pasteurization or sterilization. Steam might also increase the vulnerability of the substrate to degradation through carbohydrates (Bigelis, 1993). Additionally, to cooking or steaming, substrates like cereal grains might be crushed, milled, cracked, blended, or powdered, to fulfill the necessities of the SPF or even cooked to improve food fermentations of the flavor compounds.

Economical raw, wastes of agro-industry might be crushed or chopped to reduce the size and enhance the exposure of substrate; sanitized and/or cleansed; steamed or cooked; partly degraded through additional mixtures of chemical, physical, and/or enzymatic pretreatments; and enhanced with nutrient supplements to endorse the growth of microbes (Raimbault, 1998). Inert support included in a bed of substrate might assist in the holding of auspicious spatial along with physical features throughout microbial function on the ingredients, and develop the managing of mass transference also heat throughout SPF (Ooijkaas, 2000). Usage of inert supports might also have processing benefits throughout the recovery of the product.

4.4.3. Temperature, pH, and Moisture Control

For the creation of the product also its steadiness, the temperature optima throughout SPF might vary through culture situations, be greatly appropriate for general bacteria and fungi, which usually produce in the extent of 20–55°C. For certain SPF substrates, particularly lignocellulosic or starchy solid substrates, these optima might not be accommodated due to less water content and poor thermal conductivity. For instance, accumulation of heat might be a consequence of bacterial breakdown and could stretch to the highest stages as big as 80°C in a composting situation (Mason and Milke, 2005), through inclines of thermal spreading horizontally also vertically. Therefore, checking also regulating moisture and temperature is necessary to evade medium dryness, reduced a_w, and related reduction in accessibility of nutrients (Gutierrez et al., 1996). Heat accumulation solutions emphasize forced airing, evaporative cooling, or stirring, though the latter method risks solid materials drying and loss of moisture. The temperature is usually measured in a solid sheet, also it is examined on bioreactor entrance and departure areas. The thermosensors are generally spread out and injected radially from the fermentor middle. The sensors that are usually working are thermistors or metallic probes thermocouples. The Thermistors provide the benefit of fewer reply times, also greater sensitivity however might show lesser extended-term steadiness than metallic probes. Metallic probes, although exact and simple, might be more costly and slower to reply. Thermocouples give good reliability; however, disadvantages are related to comparative temperature measurement other than actual temperature through this method. Several sensing of temperature methods had been debated in Bellon-Maurel et al. (2003).

4.4.4. General Process for SPF

- **Culture:** Select suitable solid phase fermentation microorganism, choosing a natural, pure, or mixed culture. Develop an archival culture storing strategy. Describe the seed-making protocol and place for producing reliable inocula. Assess the distinctive requirements of bacterium, its consideration to storing, handling, and automated SPF forces, directing for reproducibility and culture steadiness.
- **Safety:** Execute a safety assessment and form biohazard regulation for SPF procedure also capacity, fulfilling the necessities of regulators and good lab functioning.
- **Pilot Studies:** Execute initial studies through a little measure of SPF to define the finest situations for development, creation of a product, and scale-up ultimately. Understand the kinetics growth in reply to variables of environment, directing the simulations explaining production; also develop an economic evaluation taking the cost of the raw materials. Confirm the product's steadiness.
- **Substrate, Solid Support, Scale, and Nutrients:** Select an appropriate nutrient and solid substrate preparation, and dense backing if valuable, founded in studies of the medium. Endorse scalability. Understand the medium physicochemical features,

for example, thermal conductivity, mixability, permeability, heat, and mass transfer, moisture content also holding, and probable variations that might happen throughout SPF, collectively with macro also micro-features of the solid substrate.

- **Pretreatment of a Solid Substrate:** Assess also appraise pretreatment procedures for solid substrate and its creation, for example., chemical or enzymatic, physical processing; pasteurization; cooking or steaming; autoclaving; blending with backing, also enhancement.
- **Inoculation:** Start the SPF through a mycelial or preparation of spore distributed through mixing/spraying utilizing a measured inoculum of suitable thickness for the speeding up of fermentation.
- **Fermentation and Checking of Parameters:** Execute the SPF below agitated or stationary rise circumstances, having a suitable bioreactor design that allows checking and procedure control on the preferred scale of environmental factors. a_w, temperature, aeration, managing moisture, pH, CO_2, O_2, mixing effectiveness, development, and association with productivity and yield. Decrease the negative impact of by-products on produce, and improve the potential coproducts value.
- **Processing and Retrieval of the Product:** Describe specifications of product and clarity requirements, and perform the recovery plan through processing phases that enhance competence and produce, recycling solvents, and materials processing whenever probable.

4.5. BIOREACTORS/FERMENTORS

Fermentors, or bioreactors, had been utilized in large and small-extent Solid Phase fermentation to generate various metabolites, enzyme products, also applied to procedures including fermented food products, biofertilizers, bioremediation, eatable mushrooms, biopesticides, and feeds (Mitchell et al., 2000). However, laboratory gauge SPF uses jars, flasks, columns, roller bottles also Petri dishes, general bioreactor/fermentor structures utilized for scaling up of industry are rotary drum fermenters, packed-bed fermenters, fluidized-bed fermentors, and tray fermenters. Mostly these structures are aerobic and frequently need aeration, either continuous or intermittent. Rare SPF procedures are anaerobic and need no aeration or agitation. Certain SPF procedures do not need bioreactors: dispersing substrate on the soil would be sufficient, also with certain types of composting and fodder fermentation (Figure 4.4) (Haug, 1993).

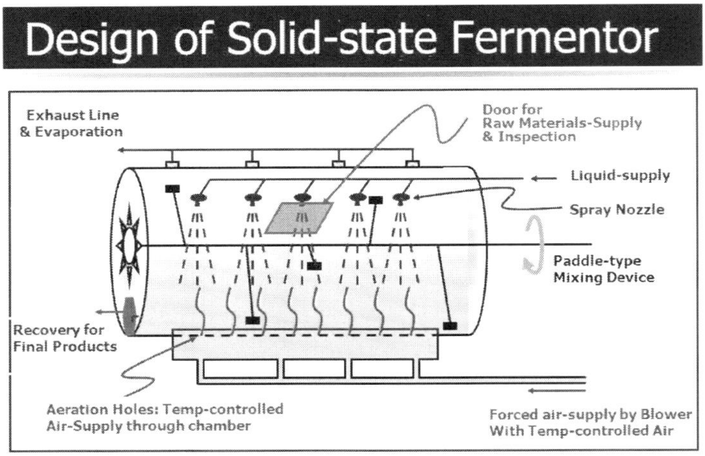

Figure 4.4. Diagram of a solid-state bioreactor.

Source: https://slideplayer.com/slide/5000523/.

4.5.1. Tray Bioreactor

Tray bioreactor comprises a pure substrate bed having no forced airing. Several trays might be fixed in a chamber through a regulated environment, where the air is dispersed around the array and comparatively humidity and temperature are controlled to maximize production and effectiveness. The exposed top and pierced bottom encourage maximum surface area for the exchange of gas. Sterilized water could be sprayed to retain levels of moisture. Physical mixing could be utilized to encourage cooling and microbial development. SPF's traditional technique using flat static trays (often on mobile trolleys) needs the share of substantial workspace and labor for huge scale production and is frequently designed for the constant process (Sato and Sudo, 1999; Krishna, 2005). Computerized handling systems had been defined, and certain had been used to the rice koji production for shochu, soy sauce, and sake (Sato and Sudo, 1999). PlaFractor™ automated technology made in Bangalore, India by Biocon is a controlled modular structure that includes mixing arms with blades, greater temperature control, sealed trays, saving a lot of energy and space.

Tray design is plain and had restrictions. The open tray is normally filled with a substrate having a depth of around 5–15 centimeters. The height of the bed cannot be enlarged because this results in overheating; therefore, the only choice for scaling-up is the enhancement in tray area or the number of trays. Mass and heat transfer restrictions, also temperature gradients, limit the application of this technique and enhance a list of drawbacks for the tray bioreactors, which also comprises ineffective use of substrate and pollution risk (Rajagopalan and Modak, 1995). Tray system is appropriate mainly for conditions with little volume requirements, and areas where there is low labor cost and technology, are unobtainable.

Tray bioreactor is historically significant as it is founded on the koji procedure, and in 1891 it was presented to the Western world by J. Takamine to produce commercial enzymes through A. oryzae grown on wheat fiber ("Taka-koji"). This kind of reactor structure was scaled up to produce citric acid in the 20th century (Figure 4.5) (Lockwood, 1979).

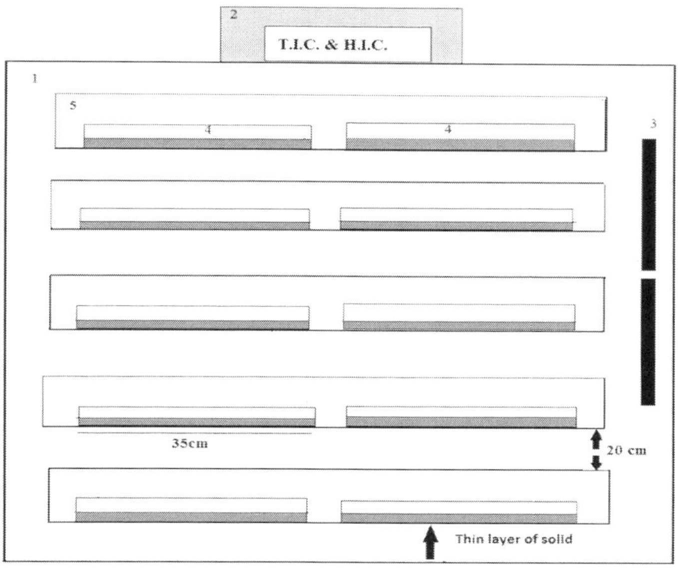

Figure 4.5. Schematic diagram of the tray bioreactor. (1) Chamber; (2) temperature and humidity indicator and controller; (3) fan, (4) tray.

Source: https://www.researchgate.net/figure/Schematic-diagram-of-the-tray-bioreactor-1-Chamber-2-Temperature-and-Humidity_fig1_341918361.

4.5.2. Packed-Bed Bioreactor

Packed-bed bioreactor normally comprises of a stationary bed in a cylinder-shaped column having forced airing by the substrate, that rests on a plate having pierced base (Mitchell et al., 2000). Alternative forms of this structure adapt a forced airing device placed in the middle of the bed for improvement of performance (Mitchell et al., 2006). The packed-bed bioreactors are hindered through the creation of axial gradients, taking to endwise airing. The rise of temperature, decrease bed porosity, had a more negative effect on performance than irregular oxygen distribution could have. Air having water saturation could improve some of the temperature impacts and effectively eradicate metabolic heat. However, this approach also raises evaporation and drying, thus microbial growth is restricted. For the circulation of water, a jacket might serve as an alternative means for regulating temperature. A specifically designed bioreactor called the Zymotis packed-bed reactor attached to eliminate heat through radial conduction, works closely with spaced inner heat transfer plates and had attained superior enzyme produces and productivities as a consequence of enhanced metabolic heat elimination (Roussos et al., 1997). Bioreactors with sporadically stirred beds look like packed-bed systems, however, they include an agitator, and are preferred for huge-scale operations, and mixing is allowed (Figure 4.6).

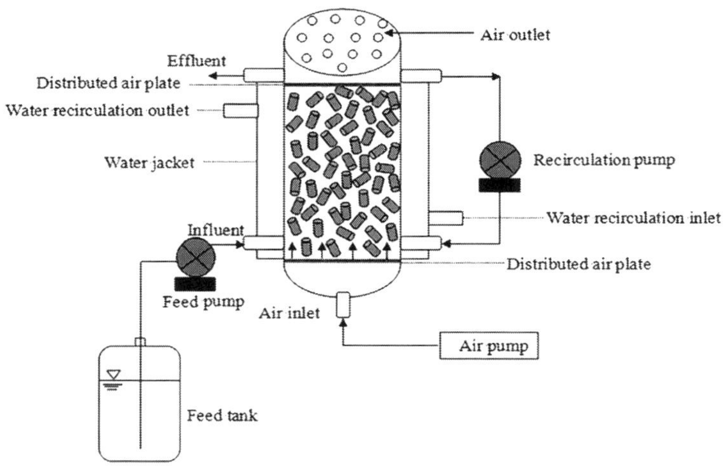

Figure 4.6. Diagram of the packed-bed bioreactor. Reactor working volume = 1.616 L.

Source: https://www.researchgate.net/figure/Schematic-diagram-of-the-packed-bed-bioreactor-Reactor-working-volume-1616-L_fig3_273513104.

The packed-bed bioreactors had restrictions further to temperature hike. Dropping of pressure (0.1–0.5-centimeter water per centimeter bioreactor) is a feature of packed-bed bioreactors that confines performance through decreasing airflow, also encouraging channeling of airflow and substrate bed cracking along with the reactor sides (Mitchell et al., 2000). Further disadvantages comprise bed caking, nonuniform development, problematic constant operation, scale-up problems, and problems in recovery of the product.

4.5.3. Rotary Drum Bioreactor

A rotary drum fermentor holds a complicated operational structure that spins a drum horizontally (or closely so), constantly, or occasionally. From the top of the substrate, the air is blown across by the headspace, distinct like the structure of the packed bed. The design of the rotating drum might include dampers to different internal lifters or airflow, baffles, or paddles, for mixing enhancement. The main drawback to this kind of bioreactor is the less fill volume; the substrate bed is not properly mixed if it is filled past 30% size. Further concerns are associated with not properly mixing (slumping flow), mass, and heat transfer, axial temperature rise, and cutoff in the substrate bed (Smith, 2007). However, the rotary drum bioreactor was utilized to produce-amylase initially in the 19th century, and in the mid of the 20th-century penicillin was produced, presently this bioreactor system is seldom used due to complex reactor construction and tough operation (Figure 4.7) (Wang and Yang, 2007).

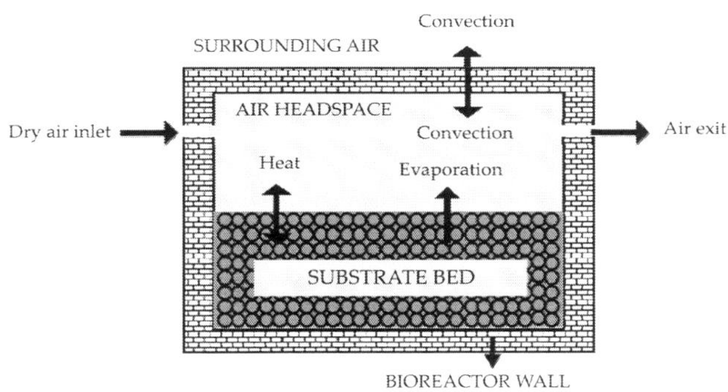

Figure 4.7. Rotating-drum bioreactor.

Source: https://www.researchgate.net/figure/Rotating-drum-bioreactor_fig1_309172326.

4.5.4. Fluidized-Bed Bioreactor

The Fluidized-bed bioreactors employed both forced airing and constant mixing (Wang and Yang, 2007). Gas is energetically directed up by a perforated base plate for fluidization of substrate particles, through fine particles it works finest. The bottom mixer might be incorporated to avoid substrate agglomeration through huge, cohesive particles, which might be tough for fluidization (Figure 4.8).

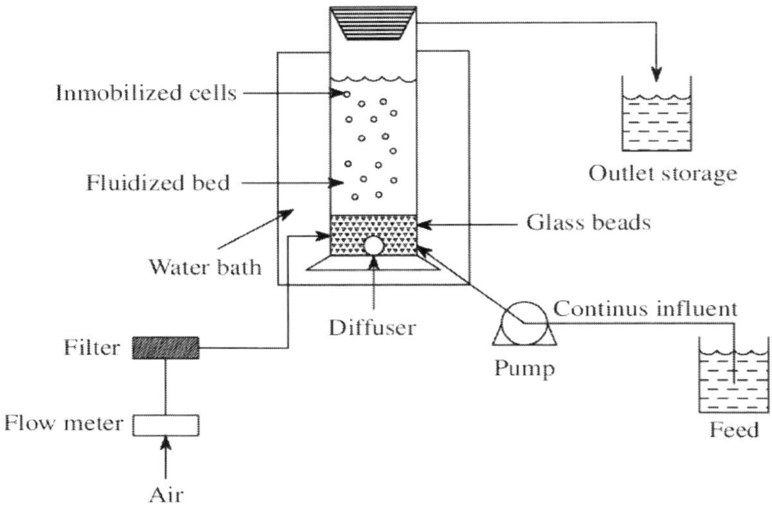

Figure 4.8. Schematic diagram of fluidized bed bioreactor structure.

Source: https://www.researchgate.net/figure/The-schematic-diagram-of-fluidized-bed-bioreactor-system_fig1_45800021.

A large flow rate of gas confirms fine mass and heat transfer amongst the gas phase and substrate particles. The structure of gas-solid fluidized had fascinated research in recent years focusing on the production of ethanol and yeast biomass (Smith, 2007). Regardless of the benefits of good mass and heat transfer, additional research is essential using these kinds of bioreactors to overwhelm the disturbance of microbial integrity and task caused because of constant agitation structures through forces of high cut-off. Furthermore, increased productivity attained with this kind of fluidized bioreactor necessarily balances high energy costs related to constant airing, also for fluidization, a high air velocity is needed (Wang and Yang, 2007).

REFERENCES

1. Acuna-Arguelles, M. E., Gutierrez-Rojas, M., Viniegra-Gonzalez, G., & Favela-Torres, E., (1995) Production and properties of three pectinolytic activities produced by Aspergillus niger in submerged and solid-state fermentation. Appl. Microbiol. Biotechnol., 43, 808–814.
2. Acuna-Arguelles, M., Gutierrez-Rojas, M., Viniegra-Gonzalez, G., & Favela-Torres, E., (1994). Effect of water activity on exo-pectinase production by Aspergillus niger CH4 on solid-state fermentation. Biotechnol. Lett., 16, 23–28.
3. Aguilar, C. N., Contreras-Esquivel, J. C., Rodriguez, R., Prado, L. A., & Loera, O., (2004). Differences in fungal enzyme productivity in submerged and solid-state cultures. Food Sci. Biotechnol., 13, 109–113.
4. Aidoo, K. E., Hendry, R., & Wood, B. J. B., (1982). Solid substrate fermentation. Adv. Appl. Microbiol., 28, 201–237.
5. Akao, T., Gomi, K., Goto, K., Okazaki, N., & Akita, O., (2002). Subtractive cloning of cDNA from Aspergillus oryzae differentially regulated between solid-state culture and liquid (submerged) culture. Curr. Genet., 41, 275–281.
6. Ali, H. K. Q., & Zulkali, M. M. D., (2011). Design aspects of bioreactors for solid-state fermentation: A review. Chemical and Biochemical Engineering Quarterly, 25(2), 255–266.
7. Anastassiadis, S., & Rehm, H. J., (2006). Citric acid production from glucose by yeast Candida oleophila ATCC 20177 under batch, continuous and repeated batch cultivation. Electronic J. Biotechnol., 9, 26–39.
8. Ashley, V. M., Mitchell, D. A., & Howes, T., (1999). Evaluating strategies for overcoming overheating problems during solid-state fermentation in packed bed bioreactors. Biochem. Eng. J., 3, 141–150.
9. Barrios-González, J., Baños, J. G., Covarrubias, A. A., & Garay-Arroyo, A., (2008). Lovastatin biosynthetic genes of Aspergillus terreus are expressed differentially in solid-state and in liquid submerged fermentation. Appl. Microbiol. Biotechnol., 79, 179–186.
10. Bigelis, R., (2010). Solid-phase fermentation: Aerobic and anaerobic. Manual of Industrial Microbiology and Biotechnology, 117–131.

11. Bigelis, R., He, H., Yang, H. Y., Chang, L. P., & Greenstein, M., (2006). Production of fungal antibiotics using polymeric solid supports in solid-state and liquid fermentation. J. Ind. Microbiol. Biotechnol., 33, 815–826.
12. Chavan, J. K., Kadam, S. S., Salunkhe, D. K., & Beuchat, L. R., (1987). Biochemistry and technology of chickpea (Cicer arietinum L.) seeds. Critical Reviews in Food Science and Nutrition, 25(2), 107–158.
13. Corey, D. R., & Schultz, P. G., (1987). Generation of a hybrid sequence-specific single-stranded deoxyribonuclease. Science, 238(4832), 1401–1403.
14. Couto, S. R., & Sanromán, M. A., (2006). Application of solid-state fermentation to food industry: A review. J. Food Eng., 76, 291–302.
15. Dalsenter, F. D. H., Viccini, G., Barga, M. C., Mitchell, D. A., & Krieger, N., (2005). A mathematical model describing the effect of temperature variations on the kinetics of microbial growth in solid-state culture. Process Biochem., 40, 801–807.
16. Demain, A. L., (1998). Induction of microbial secondary metabolism. Int. Microbiol., 1, 259–264.
17. Demain, A. L., (2000). Microbial biotechnology. Trends Biotechnol., 13, 26–31.
18. Demirci, A., Izmirlioglu, G., & Ercan, D., (2014). Fermentation and enzyme technologies in food processing. Food Processing: Principles and Applications (2nd edn., pp. 107–136). New York: Wiley.
19. Diaz, J. C. M., Rodríguez, S., Roussos, J., & Cordova, A., (2006). Lipases from the thermotolerant fungus Rhizopus homothallicus is more thermostable when produced using solid-state fermentation than liquid fermentation procedures. Enzyme Microbial Technol., 39, 1042–1050.
20. Durand, A., (2003). Bioreactor designs for solid-state fermentation. Biochemical Engineering Journal, 13(2–3), 113–125.
21. Ghildyal, N. P., Gowthaman, M. K., Rao, K. S., & Karanth, N. G., (1994). Interaction of transport resistances with biochemical reaction in packed-bed solid-state fermentors, effect of temperature gradients. Enzyme Microb. Technol., 16, 253–257.
22. Gutierrez-Rojas, M., Amar, A. H. S., Auria, R., Revah, S., & Favela, T. E., (1996). Heat transfer in citric acid production by solid-state fermentation. Process Biochem., 31, 363–369.

23. Holker, U., & Lenz, J., (2005). Solid-state fermentation—are there any biotechnological advantages? Curr. Opin. Microbiol., 8, 301–306.
24. Krishna, C., (2005). Solid-state fermentation systems: An overview. Crit. Rev. Biotechnol., 25, 1–30.
25. Larroche, D., (1996). Microbial growth and sporulation behavior in solid-state fermentation. J. Sci. Ind. Res., 55, 408–423.
26. Lockwood, L. B., (1979). Production of organic acids by fermentation. Microbial Technology, 1, 355–387.
27. Mason, I. G., & Milke, M. W., (2005). Physical modeling of the composting environment: A review: Part 1, reactor systems. Waste Manag., 25, 489–500.
28. Miall, L. M., (1978). Organic acids. Primary Products of Metabolism, Economic Microbiology, 2, 47–119.
29. Mitchell, D. A., Berovic, M., & Krieger, N., (2000). Biochemical engineering aspects of solid-state bioprocessing. Adv. Biochem. Eng. Biotechnol., 68, 61–138.
30. Mitchell, D. A., Berovič, M., Nopharatana, M., & Krieger, N., (2006). The bioreactor step of SSF: A complex interaction of phenomena. In Solid-State Fermentation Bioreactors, 1(3), 13–32.
31. Mitchell, D. A., Krieger, N., Stuart, D. M., & Pandey, A., (2000). New developments in solid-state fermentation: II. Rational approaches to the design, operation and scale-up of bioreactors. Process Biochem., 35, 1211–1225.
32. Ooijkaas, L. P., Weber, F. J., Buitelaar, R. M., Tramper, J., & Rinzema, A., (2000). Defined media and inert supports, their potential as solid-state fermentation production systems. Trends Biotechnol., 18, 356–360.
33. Pandey, A., Selvakumar, P., Soccol, C. R., & Nigam P., (1999). Solid-state fermentation for the production of industrial enzymes. Curr. Sci., 77, 149–162.
34. Pandey, A., Soccol, C. R., & Mitchell, D., (2000). New developments in solid-state fermentation, I—bioprocesses and products. Process Biochem., 35, 1153–1169.
35. Raghavarao, K. S. M. S., Ranganathan, T. V., & Karanth, N. G., (2003). some engineering aspects of solid-state fermentation. Biochem. Eng. J., 13, 127–135.

36. Raimbault, M., (1998). General and microbiological aspects of solid substrate fermentation. Electronic J. Biotechnol., 1, 1–15.
37. Rajagopalan, S., & Modak, J. M., (1995). Modeling of heat and mass transfer for solid-state fermentation process in tray bioreactor. Bioprocess Biosyst. Eng., 13, 161–169.
38. Robinson, T., Singh, D., & Nigam, P., (2001). Solid-state fermentation, a promising microbial technology for secondary metabolite production. Appl. Microbiol. Biotechnol., 55, 284–289.
39. Roussos, S., Bresson, E., Saucedo-Castaneda, G., Martinez, P., Guymberteau, J., & Olivier, J. M., (1997). Production of mycelial cells of Pleurotus opuntiae on natural support in solid-state fermentation. Advances in Solid State Fermentation, 4, 483–498.
40. Sato, K., & Sudo, S., (1999). Small-scale solid-state fermentation. Manual of Industrial Microbiology and Biotechnology, 5, 61–89.
41. Saucedo-Castaneda, G., Gutierrez-Rojas, M., Bacquet, G., Raimbault, M., & Viniegra-Gonzalez, G., (1990). Heat transfer simulation in solid substrate fermentation. Biotechnol. Bioeng., 5, 802–808.
42. Smith, P. G., (2007). Solid-state fermentation. Gas-Solid Fluidized Bed Fermentation, 1(4), 12–19.
43. Sullivan, D. M., & Miller, R. O., (2001). Compost quality attributes, measurements, and variability. Compost Utilization in Horticultural Cropping Systems, 95–120.
44. Suryanarayan, S., (2003). Current industrial practice in solid-state fermentation for secondary metabolite production, the Biocon India experience. Biochem. Eng. J., 13, 189–195.
45. Varzakas, T. H., Roussos, S., & Arvanitoyannis, I. S., (2008). Glucoamylases production of Aspergillus niger in solid-state fermentation using a continuous counter-current reactor. Int. J. Food Sci. Technol., 43, 1159–1168.
46. Wang, L., & Yang, S. T., (2007). Bioprocessing for value-added products from renewable resources. New technologies and applications. Solid-State Fermentation and its Applications, 4(12), 465–489.

Chapter 5

Scaling-Up of Industrial Microbial Processes

CONTENTS

5.1. INTRODUCTION

The scale-up and procedure improvement are designed for sustaining homogenous and ideal reaction circumstances lessening the development of metabolic precision and the revelation of microbial stress to guarantee constant item value and to increase yield, which are the main problems of industrial fermentation. A thorough and complete procedure characterization is done to explain appropriate plans for all specific facilities, items, and procedures (Gogate et al., 2011). The chosen scale-up limitations need to be kept consistent as they affect item value and yield after being recognized as the most pertinent procedure parameters. The combination of dimensionless terms, relevant parameters, or suitable mathematical collections with physical factors may only be independently kept consistent as sole limits. Instruments and methods such as scale-down simulations, computational fluid dynamics clarifying the streaming and mixing circumstances, and neural system-based chemometric information modeling and examination may help procedure characterization, which is derived from near real-time or real-time information gathered online or in-situ calculations. But the procedure circumstances may vary considerably in the culture container as accommodations for fermentation are not sternly intended to suit scale-up standards. An effective scale-up is mostly a consequence of a distinct procedure upgrade and expansion on every scale and is not the consequence of an investigational plan that is straight-lined and irrefutable since any prototypes and plans may only inadequately imitate and ponder the joint interaction and intricate extremely the interdependency of fermentation limits (Pohar and Likozar, 2014).

Stocks that are more than what growth procedure and lab microbe require are needed for scaling up manufactured microbial procedures for marketable production as it needs time and is an unrelenting attempt. The plan can be greatly affected by oversights, mistakes, and blunders, which can be expensive. It can be fruitful if handled accurately. The foundation is formed by 3 guiding values, which are: the unforeseen should be prepared for, particulars should be taken care of, and the end should be kept in sight from the start. Downstream refining is impacted by the fermentation stage, which is mostly the priciest. Samples of the suggested tactics and scale-up errors are provided. A gainful and smooth route to the large-scale marketable end will be ensured by engagement in procedure and microbe progression for the guidance of R&D, of scale-up and engineering sources accomplished in combined procedure advancement from the very start (Figure 5.1) (Binswanger et al., 2010).

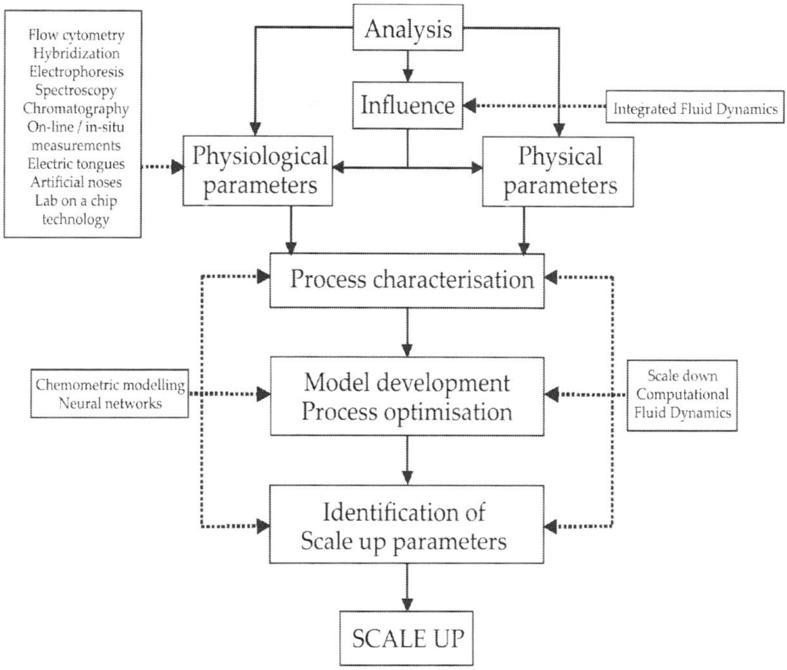

Figure 5.1. Categorized by gliding changes, arrangement generally showing the phases and theoretical stages of fermentation scale-up and procedure enhancement.

Source: https://www.semanticscholar.org/paper/Optimization-and-scale-up-of-industrial-processes-Schmidt/1b9a9f794ac3c0056c834dfc2b0d53be918e349b/figure/0.

The growth in manufacturing, size, and quantity of anything is defined as scale-up in Cambridge Dictionary. The discarding of rubbish, as well as the cleansing and following retrieval of the item after the development of microbes in bioreactors to create an item, is called microbial procedures. The generation of a monetary return to stakeholders and to give advantages to consumers is the reason the scale-up of microbial procedures is approached for a marketable purpose (Yang and Duncan, 2021). With scale variables fluctuating somewhere from 1,000 to 1,000,000, a complete industrial-scale procedure (e.g., 20,000–20,00,000 L bioreactors) should be developed from a procedure in a lab (e.g., 0.50–10.0 L bioreactors) (Figure 5.2).

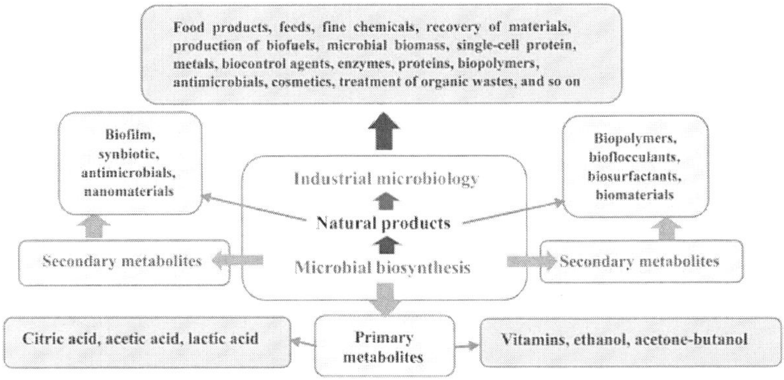

Figure 5.2. In diverse industrial areas, the significance of MSM (microbial synthetic metabolites) is shown in an organized illustration.

Source: https://www.sciencedirect.com/topics/agricultural-and-biological-sciences/industrial-microbiology.

In the case of great innovation in the marketable item or the procedure, the scale-up of procedures of large-scale industries is done in 2 phases. A coordinated downstream gear and a pilot-scale (pilot plant) with 100.0–10,000.0 L bioreactors is the initial phase. Its work is to change the manufacturing procedure to an actual scaled-down adaptation from the lab-scale procedure. The operation of every sole unit is done group-wise, i.e., mostly, the procedure is not completely combined. The needed item prototype sizes and scaled-down gear prices, as well as the accessibility and size, are the basis of judgment for the designated pilot plant (Goddard et al., 2010). A coordinated downstream gear and a demo scale (demonstration plant) with 10,000–100,000 L bioreactors is the secondary phase. The marketplace wants, the procedure, and the supply network (marketable item application from raw supplies) are validated to lessen the danger of a high capital venture in the full-scale industrial plant. Along with recycling streams, the demonstration procedure is continuous. The demo plant is omitted if the extent of innovation is small. Regarding demo and pilot scales, the term "pilot" will be employed from now on.

The price required to create a lab-scale procedure and development microbe is less than the price needed for the scale-up to industrial scale from a microbial procedure. The commencement of the industrial plant, its making, and the preliminary procedure's validation (demo and pilot scales) are included in the US $100.0 million to $ 1 billion needed for it

(Davison and Lievense, 2016). The exact sequence is followed by the yearly functional price of the plant. A period of 3–10 years is needed for the change to an industrial plant from a lab-scale one. Scheme closure may be the result as at the time of scale-up any disruptions may be pricey as the monetary risk is great under such a situation. Consumer and investor confidence will be undermined and stockholders will receive a less monetary return if postponements (3 to 12 months) happen or gradual under-performance (5.0–10.0%) at the time of scale-up takes place. It is important to get the microbial procedure correct and to get it correct the 1st time (Figure 5.3).

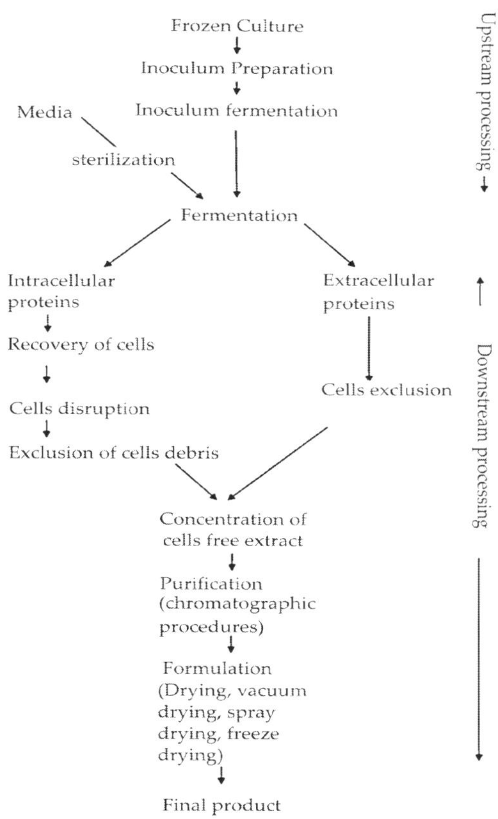

Figure 5.3. Enzymes' industrial manufacturing.

Source: https://link.springer.com/article/10.1007/s13205-016-0485-8.

5.2. PROCEDURE SCALE-UP'S GUIDING VALUES

Involvement in the popularization of various manufacturing microbial procedures such as scale-up to manufacturing sustenance from initial phase R&D, and 1st of type schemes is done by the following authors: Kennedy (2017); and Chem Process (2018). For the fruitful scale-up of manufacturing microbial procedures, three guiding values that are quite significant are mentioned in subsections (Figure 5.4).

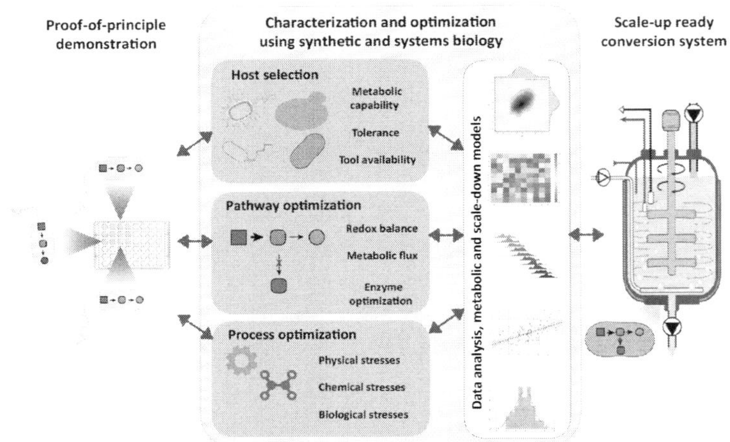

Figure 5.4. For industrialized scale-up, an enhanced conversion scheme from a proof-of-principle strain progress's system.

Source: https://www.cell.com/trends/microbiology/fulltext/S0966-842X(19)30019-8.

5.2.1. Start by Keeping the Conclusion Insight

Irrespective of the attempt, why wouldn't you? Especially when it is a 1st type of procedure, it is important to have an actual and precise view of what the conclusion will be like. It is not possible to just expand lab-scale circumstances and lab-scale apparatus to large-scale (Weiss, 2016; Noorman and Heijnen, 2017). A scheme will probably face great trouble if one does not know how large-scale reliant limitations vary and of large-scale apparatus.

Even before the initial lab experimentations are performed, a knowledgeable scheme team that starts with the conclusion in view and comprehends large-scale procedures will prepare a thorough conceptual plan of the projected industrial plant and procedure. Procedure flow illustrations, unit function models, energy, and substance balances, and techno-economic prototypes are built by the team based on actual engineering, biology, and chemistry conventions. This is the cleverest venture one can make even though its investment is insignificant in comparison to the whole scheme price (Antle, 1999). The chief scale, economic limitations, and procedure capability is guided on to the experimentation R&D agenda by using the theoretical strategy. As the production of fresh learning takes place, revise the design.

5.2.2. Attention to Details

The outcomes fluctuating from troublesome to disastrous have been the consequence of timesaving methods and omissions at the time of scale-up. Less hostile shocks will be faced if a lot of attention is given to particulars so that small-scale scale-up can take place (Matsen, 1996). An industrial plant that is dependable, harmless, and surpasses the monetary aims will be the prize for stockholders. How to correctly approach procedure scale-up and a few usual errors are highlighted in Table 5.1.

Table 5.1. Typical mistakes during scale-up and advised actions

Mistake	Suggested Approach
Plant functions were not dependable	The utility effect should be comprehended; utilities are validated before starting and utility redundancy should be engineered where possible
Goals were not reached and functioning surprisingly digressed	Intense scale-up limitations should be observed using prototypes of large-scale and evaluations should be done in lab scale-down tests often
The Fermentation network's cleanliness verification was not done	The complete sterile boundary should be assessed for which proper cleanliness tests should be done
Postponements were faced due to loss of apparatus as a program for precautionary maintenance was not in place	Equipment should be identified and skilled engineers who can place a maintenance program in place should be hired
After construction, enough money was not present	Enough money should be saved up in case of any issues and for proper beginning

Before starting, raw supplies were not authorized	Before buying, confirm all the authorized raw supplies
The conductance of 1st of a type procedure was not done	Pilot information should be used to model and run the plant and the procedure should be piloted
Product was lost as the function team did not know how to cope with upsets	A scheme in case of procedure upsets should be made and the team should be made to practice in the plant by using procedure simulators
A plant like this was never made by the construction group and engineers	Engineers and a construction group that has done a project like this before should be hired
Scheme's schedule and the price was given unbelievable limits by the technical and business teams	Defy burdens to overpromise; strain test schemes with field experts; recognize and consider plan performance dangers against prizes and drawbacks
Before the plant starts, the functioning team was not trained	In the pilot plant, the team should be trained and given experience
Delays were faced due to bad plan management	Activities should be coordinated and overseen by a skilled manager
The transfer of technology could not be done as there was no system in place and deviations in the plan could not be performed troubleshooting on.	Bioreactors should be installed for troubleshooting and transfer of tech
Intricate apparatus and instruments were not bought to save price	For procedure controlling and maintenance, proper apparatus should be recognized which is crucial to it

5.2.3. Preparation for the Unexpected

Problems will surface at the time of scale-up no matter how well one is prepared. Apparatus letdown, microbial pollution, soiling of procedure gear, utility disruptions, and unforeseen poor procedure functioning are some typical instances. This is the point where relief plans and risk evaluation pay off. Think about all the possible dangers that may take place with your team, the likelihood of every risk-taking place, and the severity of its occurrence. A thorough relief scheme should be planned and risk scale should be prioritized. The assessment of the effect of great-scale risks on all procedures should be done, and a response scheme should be created which should be relayed to all members so they can take the appropriate steps if an upset does happen (Martinez, 2010–2018).

5.3. MAIN FACTORS FOR FERMENTATION SCALE-UP

A significant scale-up hazard at large-scale is nearly always a bad fermentation operation. Keeping in view the money investment and flexible prices (services and raw ingredients), fermentation is mostly the priciest procedure stage. The nature/quantity of procedure wastes and the downstream part of procedures' functioning is impacted by it.

An intricate element operation is a fermentation (Yang, 2010). At the time of scale-up, various limitations are subject to alteration, and they also affect the functioning. Scale reliant functioning limitations are described in Table 5.2. They can harm the whole economics of the plant and procedure performance, and they alter due to gear scale and model, or due to budget limits.

Fermentation as an upgrade peril can be reduced or eliminated by paying attention to and recognizing scale-reliant fermentation limitations. Decent downstream functioning and hopes regarding steady large-scale fermentation functioning are met when a methodical attitude is taken towards it. Because of a long time of functioning experience in nonstop plant studying, a scalable and well-modeled fermentation is capable of performing better in the industrial plant than in the laboratory.

Table 5.2. Scale-dependent fermentation limitations

Parameter	Divergence/ Deviation	Effect/ Impact
Bioreactor mixing period	Magnitude intensification	Fermentation functionality can be affected negatively by intricate procedure control limitation's gradients i.e., pH, nutrient or substrate concentration, temperature
Broth hydrostatic pressure	magnitude increase along with gradient alongside the vertical axis	Fermentation functionality can be affected by partial pressure gradients, gases that are dissolved, and raised gas partial forces i.e., pCO_2, PO_2.
Broth deactivation	Temperature, constant deactivation, and residence time profile v.s batch	WWT (wastewater treatment) and DSP (downstream processing) can be negatively affected by broth chemistry and cell lysis which are affected by temperature or time i.e., deactivation situations

Raw material grade	Concentration, reagent grade, lot-to-lot- fluctuation purity vs. industrial	For fed-batch procedures, time course dynamics and water balance can be affected by differences in concentration; fermentation can be harmfully affected by components that are not fermentable and inhibitors accumulation
Broth handling	Times for harvest and broth holds being prolonged	WWT and DSP can be negatively affected by broth chemistry and cell lysis which are affected by residence circumstances (concentration of item, aerobicity, and temperature) and time period.
Cleansing of raw matter	Temperature, constant cleansing, and residence time profile vs. batch	DSP, WWT, and fermentation can be badly affected by the development of inhibitors or ruination of constituents
Liquid-gas volumetric mass transference coefficient (k_La)	Apparatus modeling causing upper restriction; power deterioration causing gradient	Fermentation functionality can be badly affected by gradients and restrictions in mass transference degrees
Shear stress	Magnitude amplification	DSP and fermentation functionality can be affected by cell impairment caused by great shear stress

5.4. A FRUITFUL SCALE-UP'S GROUND PLAN

One should start with the conclusion in view so that the required conclusion should be envisioned before the journey begins of a vigorous complete-scale industrial plant that will reach its marketable goals, i.e., price, value, and time. As the plan moves towards procedure advancement and placement from R&D proof-of-concept, the inscribed charter which is made at the start of the scheme in the shape of a comprehensive charter is updated (Borenstein and Koren, 1991).

The envisioned large-scale industrial procedure should be mimicked as much as it can during the conductance of lab-scale procedure advancement. The procedure is possible to be scaled up properly if the procedure is scaled down in a proper way (Yang, 2010; Noorman and Heijnen, 2017). As typical small-scale apparatus is not obtainable, some procedure element functions face restrictions in the scale down. By implementing a few customized regulated software and hardware, large-scale settings can mostly be sufficiently replicated in stirred lab bioreactors in fermentation's case. In intense scale-up limitations, vibratory behavior can be elicited by customizing regulated algorithms.

To imitate gas stage heterogeneity, active vibrations in movement degree ratio and great gas fractional pressure are mixed with enhanced gases. For the imitation of substrate concentration gradients, the vibrating or throbbing substrate feed degree, to impersonate k_La heterogeneity vibrating agitation degree are considered, these are some instances. Fermentation procedure kinetics and fermentor hydrodynamics should be combined by large-scale bioreactor prototypes which determine periods, extents, and circumstances for testing (Bisgaard et al., 2020).

Dreadful conditions at the time of scale-up can be reduced if a lab procedure is correctly scaled down. However, the requirement of a mid-way scale procedure confirmation is not eliminated due to it, especially for 1st of type procedure. Risk can further be reduced and unique advantages are provided by piloting at 1% to 10% of the complete processing scale (Carvalho, 2011):

- The evaluation of dealers and substitute gear models can be done.
- Consumer demand for market production, connection, and confidence is built if a great number of items can be made for consumer assessment in the applications.
- Comprising all ingredients and manufacturing apparatus, a completely combined procedure can be functioned for a long time counting recycle streams.
- The upcoming large-scale plant functioning group will know how to deal with procedure upsets and have knowledge of how it works if they are trained.
- The scheme of the large-scale plant can be enhanced by functioning knowledge and pilot plant information.

Because of the problems that come up at the time of the plan, the apparatus needed and if an institution is made or has to be created because piloting needs six months to three years. From the complete scheme of price, 5% to 20% is required by piloting. This makes it attractive to pilot only a few element functions for fewer periods or to eliminate this stage. However, the advantages weigh less than the disadvantages and experience suggests this to be a foolish decision (Staffell et al., 2019).

On the basis of the figures produced in the pilot plant, the model of the large-scale industrial plant should be made. The upcoming technology enhancements should also be included in the mindset. But a procedure that has not yet been confirmed at the pilot scale should not be designed for the large-scale factory.

All stages of the scheme should be given proper technical help comprising of building, beginning, engineering model, and contracting no matter if one warrants his technology or makes, owns, and functions his factory (Petersen, 2003). The installation of intricate instruments for procedure analytics and regulation must be done for an easy factory starting. For procedure troubleshoot help and procedure transmission confirmation, the satellite laboratory apparatus is very significant. An intense response should be given to the heads about technology functioning at the time of start-up by experts of the procedure who are present on site. For issues that need intense troubleshooting, a distinct group of tech experts needs to be available throughout the day and night.

5.5. MAIN TRIALS AND ENHANCEMENT PROSPECTS

Following the three chief rules may be difficult. Various R&D schemes have a wrong or inadequate view of the market end conclusion and do not start with the outcome in view. Without considering the appropriateness of microbial strain host for large-scale development and choosing it nevertheless is a typical basic mistake. Why is that? Businessmen who do not have proper tech commercialization knowledge, engineers, and lab scientists consider fresh chances usually. How to get to the end or how it looks like it is not known to them (Köster, 2003). They might think that marketable engineering participation is a thing to be paid attention to later even though it is significant. Such schemes face problems at the time of scale-up for sure, and this approach is a huge error.

Scale-up being executed at a great level is difficult even with an open-minded approach. The mysterious, skillsets, non-idealities in humanoid sources and services are jam-packed in the actual world. An underrated feature in a lot of circumstances has been noticed to be trained scheme organization. The perfect scheme organization at the time of scale-up is not considered even by R&D scheme heads, engineers, or scientists. Being a first-rate scheme manager does not mean one is a first-rate tech professional. However, they consist of the qualities of being adaptive, systematized, integrative, knowledgeable, devoted, and open. The end is clear to them. They prepare for issues but plan for achievement (Aston, 2017). The 3 main rules are embraced and embodied by them.

It is good to learn from the ones who have done something before you, is the meaning of a very common saying, i.e., "the early bird may get the worm, but the second mouse gets the cheese." People who wake up early are some of the cleverest people we know. Even if they have never done a task before, they are sure that they can do it well than others, in a cheap way, and can do it quickly and alone. They are devoted and amazing (Colwell et al., 1987). The difficulties of scale-up are ignored or even underestimated by them. They will achieve the destiny of the initial mouse even though they may sometimes get the worm. It is better to be the next mouse. Fresh chances should be evaluated by using the experienced engineering sources around oneself, realistic hopes should be set, advice should be given to in-progress R&D and to scale up your procedure with success (Griliches, 1979).

REFERENCES

1. Aarons, G. A., Hurlburt, M., & Horwitz, S. M., (2011). Advancing a conceptual model of evidence-based practice implementation in public service sectors. Administration and Policy in Mental Health and Mental Health Services Research, 38(1), 4–23.
2. Alonso-Rasgado, T., Thompson, G., & Elfström, B. O., (2004). The design of functional (total care) products. Journal of Engineering Design, 15(6), 515–540.
3. Antle, J. M., (1999). Benefits and costs of food safety regulation. Food Policy, 24(6), 605–623.
4. Aston, W. J., Hope, D. E., Nowak, A. K., Robinson, B. W., Lake, R. A., & Lesterhuis, W. J., (2017). A systematic investigation of the maximum tolerated dose of cytotoxic chemotherapy with and without supportive care in mice. BMC Cancer, 17(1), 1–10.
5. Binswanger-Mkhize, H., & McCalla, A. F., (2010). The changing context and prospects for agricultural and rural development in Africa. Handbook of Agricultural Economics, 4, 3571–3712.
6. Bisgaard, J., Muldbak, M., Cornelissen, S., Tajsoleiman, T., Huusom, J. K., Rasmussen, T., & Gernaey, K. V., (2020). Flow-following sensor devices: A tool for bridging data and model predictions in large-scale fermentations. Computational and Structural Biotechnology Journal, 18, 2908.
7. Blum, R. W., Garell, D., Hodgman, C. H., Jorissen, T. W., Okinow, N. A., Orr, D. P., & Slap, G. B., (1993). Transition from child-centered to adult health-care systems for adolescents with chronic conditions: A position paper of the society for adolescent medicine. Journal of Adolescent Health, 14(7), 570–576.
8. Borenstein, J., & Koren, Y., (1991). The vector field histogram-fast obstacle avoidance for mobile robots. IEEE Transactions on Robotics and Automation, 7(3), 278–288.
9. Carvalho, P., (2010). Governance and Comparative Performance of Iberian Peninsula Seaports: An Application of Non-Parametric Techniques (pp. 1000–1021). Governance and Comparative Performance of Iberian Peninsula Seaports.
10. Colwell, R. P., Nix, R. P., O'donnell, J. J., Papworth, D. B., & Rodman, P. K., (1987). A VLIW architecture for a trace scheduling compiler. ACM SIGARCH Computer Architecture News, 15(5), 180–192.

11. Crater, J. S., & Lievense, J. C., (2018). Scale-up of industrial microbial processes. FEMS Microbiology Letters, 365(13), 138.
12. Crater, J. S., & Lievense, J. C., (2018). Scale-up of industrial microbial processes. FEMS Microbiology Letters, 365(13), fny138.
13. Culler, S., (2016). A bioengineering platform to industrialize biotechnology. Chem. Eng. Prog., 112, 42–51.
14. Daugulis, A. J., Tomei, M. C., & Guieysse, B., (2011). Overcoming substrate inhibition during biological treatment of monoaromatics: Recent advances in bioprocess design. Applied Microbiology and Biotechnology, 90(5), 1589–1608.
15. David, P. A., Hall, B. H., & Toole, A. A., (2000). Is public R&D a complement or substitute for private R&D? A review of the econometric evidence. Research Policy, 29(4–5), 497–529.
16. Davison, B. H., & Lievense, J., (2016). Technology challenges and opportunities in commercializing industrial biotechnology. Chemical Engineering Progress, 112(6), 1–5.
17. Garousi, V., Eskandar, M. M., & Herkiloğlu, K., (2017). Industry–academia collaborations in software testing: experience and success stories from Canada and turkey. Software Quality Journal, 25(4), 1091–1143.
18. Ge, W., & Li, J., (2003). Macro-scale phenomena reproduced in microscopic systems—pseudo-particle modeling of fluidization. Chemical Engineering Science, 58(8), 1565–1585.
19. Goddard, L., Aitchellouche, Y., Baethgen, W., Dettinger, M., Graham, R., Hayman, P., & Meinke, H., (2010). Providing seasonal-to-interannual climate information for risk management and decision-making. Procedia Environmental Sciences, 1, 81–101.
20. Gogate, P. R., Sutkar, V. S., & Pandit, A. B., (2011). Sonochemical reactors: Important design and scale up considerations with a special emphasis on heterogeneous systems. Chemical Engineering Journal, 166(3), 1066–1082.
21. Griliches, Z., (1979). Issues in assessing the contribution of research and development to productivity growth. The Bell Journal of Economics, 92–116.
22. Howell, D. J., (1979). Flock foraging in nectar-feeding bats: Advantages to the bats and to the host plants. The American Naturalist, 114(1), 23–49.

23. Jovanov, E., Milenkovic, A., Otto, C., & De Groen, P. C., (2005). A wireless body area network of intelligent motion sensors for computer assisted physical rehabilitation. Journal of NeuroEngineering and Rehabilitation, 2(1), 1–10.
24. Knorr-Cetina, K. D., (1982). Scientific communities or transepistemic arenas of research? A critique of quasi-economic models of science. Social Studies of Science, 12(1), 101–130.
25. Köster, E. P., (2003). The psychology of food choice: Some often encountered fallacies. Food Quality and Preference, 14(5–6), 359–373.
26. Lockwood, J. A., Bomar, C. R., & Ewen, A. B., (1999). The history of biological control with Nosema locustae: Lessons for locust management. International Journal of Tropical Insect Science, 19(4), 333–350.
27. Martin, Y., & Vermette, P., (2005). Bioreactors for tissue mass culture: design, characterization, and recent advances. Biomaterials, 26(35), 7481–7503.
28. Matsen, J. M., (1996). Scale-up of fluidized bed processes: Principle and practice. Powder Technology, 88(3), 237–244.
29. Musinguzi, M., Enemark, S., & Mwesigye, S. P., (2021). Fit for purpose land administration: Country implementation strategy for addressing Uganda's land tenure security problems. Land, 10(6), 629.
30. Noorman, H. J., & Heijnen, J. J., (2017). Biochemical engineering's grand adventure. Chemical Engineering Science, 170, 677–693.
31. O'Hara, E., Neves, A. L., Song, Y., & Guan, L. L., (2020). The role of the gut microbiome in cattle production and health: driver or passenger? Annual Review of Animal Biosciences, 8, 199–220.
32. Ondrey, G., (2013). 42nd Kirkpatrick award announced. Chemical Engineering, 120(11), 15.
33. Patnaik, P. R., (1999). Neural control of an imperfectly mixed fed-batch bioreactor for recombinant β-galactosidase. Biochemical Engineering Journal, 3(2), 113–120.
34. Petersen, P. E., & Yamamoto, T., (2005). Improving the oral health of older people: The approach of the WHO Global Oral Health Program. Community Dentistry and Oral Epidemiology, 33(2), 81–92.
35. Petersen, P. E., (2003). The World Oral Health Report-2003: Continuous improvement of oral health in the 21st century–the approach of the

WHO Global Oral Health Program. Community Dentistry and Oral Epidemiology, 31, 3–24.

36. Petersen, P. E., (2004). Challenges to improvement of oral health in the 21st century—the approach of the WHO Global Oral Health Program. International Dental Journal, 54, 329–343.
37. Petersen, P. E., (2004). Improvement of oral health in Africa in the 21st century the role of the WHO Global Oral Health Program. African Journal of Oral Health, 1(1), 2–16.
38. Pohar, A., & Likozar, B., (2014). Dissolution, nucleation, crystal growth, crystal aggregation, and particle breakage of amlodipine salts: Modeling crystallization kinetics and thermodynamic equilibrium, scale-up, and optimization. Industrial and Engineering Chemistry Research, 53(26), 10762–10774.
39. Rabin, M., (1998). Psychology and economics. Journal of Economic Literature, 36(1), 11–46.
40. Staffell, I., Scamman, D., Abad, A. V., Balcombe, P., Dodds, P. E., Ekins, P., & Ward, K. R., (2019). The role of hydrogen and fuel cells in the global energy system. Energy and Environmental Science, 12(2), 463–491.
41. Sun, J., & Yan, Y., (2016). Non-intrusive measurement and hydrodynamics characterization of gas-solid fluidized beds: A review. Measurement Science and Technology, 27(11), 112001.
42. Thiering, R., Dehghani, F., & Foster, N. R., (2001). Current issues relating to anti-solvent micronization techniques and their extension to industrial scales. The Journal of Supercritical Fluids, 21(2), 159–177.
43. Tushman, M., & Nadler, D., (1986). Organizing for innovation. California Management Review, 28(3), 74–92.
44. Wiltbank, R., Read, S., Dew, N., & Sarasvathy, S. D., (2009). Prediction and control under uncertainty: Outcomes in angel investing. Journal of Business Venturing, 24(2), 116–133.
45. Zhang, Z., (1999). Flexible camera calibration by viewing a plane from unknown orientations. In: Proceedings of the Seventh IEEE International Conference on Computer Vision (Vol. 1, pp. 666–673).

Chapter

6

Dietary Uses of Microbial Fermentation

CONTENTS

6.1. INTRODUCTION

The utilization of different kinds of microorganisms in the preparation of various foodstuff is an ancient method. Technological advancements in developing innovative enzymes having broad applications range with specificity have opened new doors to explore further novel application areas. Various microorganisms comprising yeast, bacteria, and fungi as well as their enzymes have been extensively utilized in multiple foods for improvement in the taste and enhancement in their texture, providing enormous monetary paybacks to the food industry. Moreover, microorganism-based enzymes are the ideal source for plants or animals because of numerous advantages comprising cost-effectiveness, easy, and steady production. The present review deliberates the latest advancements in the development of enzymes for food-based industrial applications. An inclusive list of enzymes utilized in the processing of various foods, their microbial source, and their applications in various industries are conferred.

Fermentation of food through microorganisms has been carried out since the distant past and is still being used for the preparation of various food items (Soccol et al., 2006). Microorganisms-based enzymes perform a foremost role in the food industry due to their high stability compared to enzymes obtained from different plants and animals. Microbial enzymes can be acquired by various fermentation methods in a low-cost way with reduced requirements of time as well as space, and owing to their elevated constancy, process optimization and alteration can be accomplished without difficulty (Figure 6.1) (Kohlhepp, 2010).

Different kinds of enzymes are used in various industries for numerous applications in various industries, such as amylolytic enzymes are extensively used in the food industry, paper industry, textile industry, and detergent industry (Pandey et al., 2000). These enzymes are utilized to produce syrups of glucose, maltose, high fructose corn, and crystalline glucose, etc. In detergent industries, they have usages as additives for eradicating starch-based stains. These enzymes are utilized for decreasing the viscosity of starch in numerous paper industries for achieving proper coating on paper. Furthermore, in textile industries, amylases have been used to acquire warp sizing for textile fibers (Magalhães and Souza, 2010).

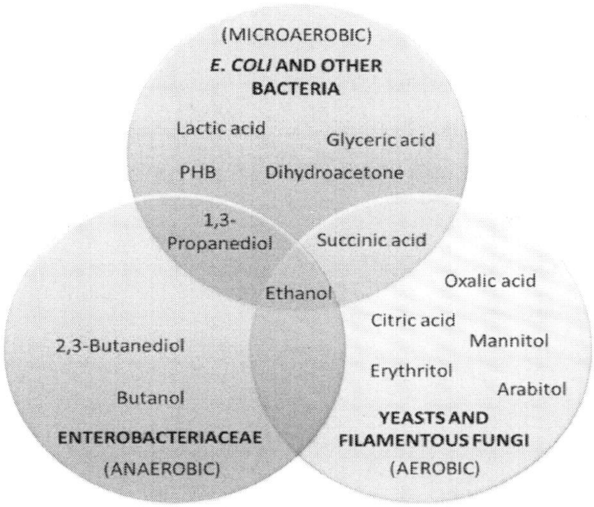

Figure 6.1. Various chemicals produced in crude glycerol by microbial fermentation. Circles/positions specify the aerobic conditions in which these chemicals are created through fermentation and the main microbial producing groups.

Source: https://www.researchgate.net/figure/Examples-of-chemicals-produced-by-microbial-fermentation-of-crude-glycerol_fig2_229324942.

Likewise, proteases enzymes, lipases enzymes, and xylanases enzymes possess an inclusive range of applications in the food industry. The succeeding units provide comprehensive and reorganized info regarding several food enzymes of the microbial source. Table 6.1 provides an outline of applications related to microorganisms-based enzymes in food industries (Couto and Sanromán, 2005; Blanco et al., 2014).

Table 6.1. Various Applications of Food Enzymes of Microbial Origin

SL. No.	Microbial Enzyme	Application	References
1.	Glucose oxidase enzyme	Improvement in shelf life and flavor of food	Hanft and Koehler (2006)
2.	Catalase enzyme	Preservation of food (glucose oxidase) Hydrogen peroxide removal from milk before cheese manufacturing	Ough (1975)
3.	Glucoamylase enzyme	Beer making Bread quality enhancement Production of high fructose and glucose syrups	Blanco et al. (2014)
4.	α-Acetolactate dehydrogenase enzyme	Decreasing beer maturation	Kumar et al. (2014)
5.	Protease enzyme	Beer making Tenderization of meat Thickening of milk Enhancing quality of bread	Aruna et al. (2014)
6.	α-Amylase enzyme	Baking, brewing, liquefaction of starch Enhancing quality of bread Rice cakes Fruit juice clarification	Couto and Sanromán (2005)
7.	Peroxidase enzyme	Improvement in flavor, color, and enhancement in the nutritional quality of food	Regalado et al. (2004)
8.	Lactase (β-galactosidase) enzyme	Lactose intolerance drop in humans Prebiotic food constituents	Jurado et al. (2002)
9.	Xylanase enzyme	Clarification of fruit juice Enhancement in beer quality	Camacho and Aguilar (2003)
10.	Lipase enzyme	Cheese flavor development Cheddar cheese manufacture	Jooyandeh et al. (2009)
11.	Cellulase enzyme	Animal feeds Fruit juice clarification	Mekala et al. (2008)
12.	Phospholipase enzyme	Development of cheese flavor Lipolyzed milk fat production	Law (2010)
13.	Debittering enzymes-naringinase	Elimination of unpleasant flavor in fruit juice Wine aroma enrichment	Zhu et al. (2017)
14.	Esterase enzyme	Flavor and fragrance improvement in fruit juice De-esterification of dietary fiber Short-chain flavor esters production	Panda and Gowrishankar (2005)

15.	Asparaginase enzyme	Reducing the formation of acryl-amide during baking	Kumar et al. (2014)
16.	Laccase enzyme	Polyphenol elimination from wine Baking	Tanrıöven and Ekşi (2005)
17.	Pectinase enzyme	Clarification of fruit juice	Pasha et al. (2013)

6.2. 1A-AMYLASE ENZYME

α-Amylases enzymes are starch-degrading and can hydrolyze α-1,4 glycosidic bonds of polysaccharides, resulting in short-chain dextrins creation. All living organisms possess these enzymes. α-amylase enzymes are mostly metalloenzymes and need calcium ions to sustain associated activity, integrity, and stability (Figure 6.2) (Raveendran et al., 2018).

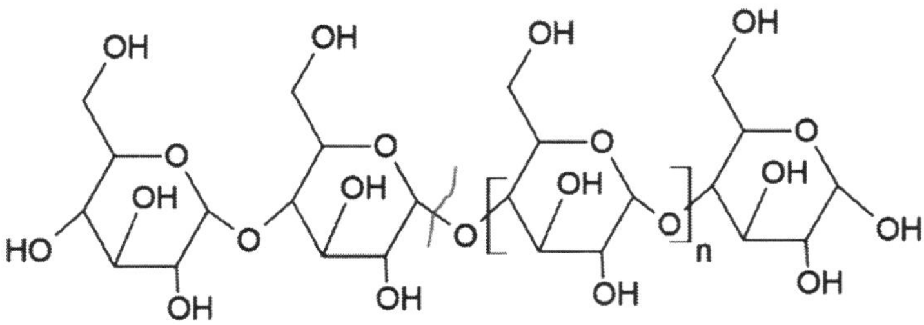

Figure 6.2. α-Amylase structure.

Source: https://www.aaltoscientific.com/product/amylase/.

α-amylases have widespread applications in food industries comprising baking, starch liquefaction, brewing, and as a digestive aid (Couto and Sanromán, 2005). These enzymes have extensive use in the baking industry to improve flavors as well as an antistaling agent for improving the quality of bread. In the course of baking, α-amylases enzymes are mixed in the dough to convert starch into small dextrins, later fermented by yeast. These enzymes improve the color of crust, taste, and toasting qualities while preparing bread in baking industries. α-Amylases have also been used for manufacturing high-molecular-mass branched dextrins. In addition, these enzymes are also utilized as a glazing agent to produce powdery foods and rice cakes (Blanco et al., 2014). In the starch industries, they are used for liquefaction of starch to convert starch into fructose and glucose syrups. Microbial enzymatic transformation of starch implicates three stages, i.e., gelatinization, followed by liquefaction, and later saccharification. During gelatinization, the viscous suspension is formed by dissolving starch granules. The second step entails the liquefaction process, in which viscosity is reduced by partial hydrolysis. In the last step, maltose and glucose are produced by the saccharification process. This process needs exceedingly thermally stable enzymes, and mostly the saccharification of starch is achieved by using α-amylases associated to Bacillus amyloliquefaciencs, Bacillus licheniformis, or Bacillus stearothermophilus (Magalhães and Souza, 2010). The ethanol production involves starch conversion to fermentable sugars through α-amylase enzymes activity and later fermentation of the sugars is attained by Saccharomyces cerevisiae to produce alcohol (ethanol). Further usages of α-amylases comprise clarification of fruit juices in the existence of pectinases and cellulases for improved yield and to make the fermentation process lucrative.

6.3. GLUCOAMYLASE (GA) ENZYME

These enzymes are exo-acting catalyzing the hydrolysis process of polysaccharide starch at non-reducing end, liberating β-glucose. Glucoamylases (GA) enzymes are also termed as saccharifying enzymes, which are broadly distributed among all living creatures. GA is acquired primarily from Aspergillus awamori and Aspergillus niger, but the enzyme obtained from Rhizopus oryzae is extensively utilized in numerous industries (Figure 6.3) (Coutinho and Reilly, 1997).

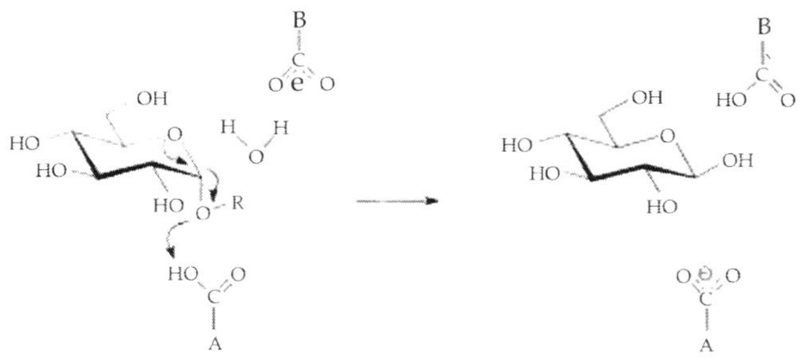

Figure 6.3. The structure of glucoamylase (GA) and its catalytic mechanism demonstrating the action of the catalytic base E400 (left) and acid E179 (right) in the water-assisted hydrolysis of the substrate comprising inversion of the configuration of the anomeric carbon.

Source: https://www.sciencedirect.com/science/article/abs/pii/S0167483800002326.

GA are mostly stable at low temperatures. However, at elevated temperatures, these enzymes drop their activity because of conformational change. GA has a wide range of applications in various food industries, including the production of high fructose and glucose syrups. In the baking industry, they are used for improvement in quality of flour, reduction of dough staling, improvement in bread crust color, and enhancing the quality of baked products of high fibers. They are used to transform the starch existing in the flour into fermentable sugars and maltose, later fermented by the yeast to rise the dough. They are also used to yield glucose, and after fermentation with Saccharomyces cerevisiae produces ethanol. GAs perform a significant role in making soya sauce, sake, and light beer. They are used to metabolize dextrins for transforming them into fermentable sugars with curtailed value of calories and alcohol in beer (Blanco et al., 2014).

6.4. PROTEASE ENZYME

Proteases are used for catalyzing the hydrolysis of peptide bonds existing in polypeptides and proteins. These enzymes have been extensively utilized in the pharmaceutical, detergent, and food industries. These enzymes occupy 60% of the market of industrially utilized enzymes. The worldwide requirement for protease enzymes was increased at a compound annual growth rate (CAGR) of 5.3% during 2014–2019. The requirement of these enzymes is anticipated to be increased further due to their applications in bioremediation processes and the leather processing industry. Protease enzymes are categorized based on their source, catalytic activity, and the type of reactive group in the catalytic site. Microorganisms (bacteria and fungi), animals, and plants are primary sources of protease enzymes. Protease enzymes are further classified in two major groups, comprising, exopeptidases, and endopeptidases, according to their catalytic sites on polypeptide chains (Law, 2010). The exopeptidases perform the activity at the edges of polypeptide chains, whereas, endopeptidases arbitrarily act at the inner sites of the polypeptide chains, leading to additional classification of endopeptidases in six groups, of the catalytic residue existent in the active site: cysteine, metallo, serine, glutamic acid, threonine protease, and aspartic (Figure 6.4).

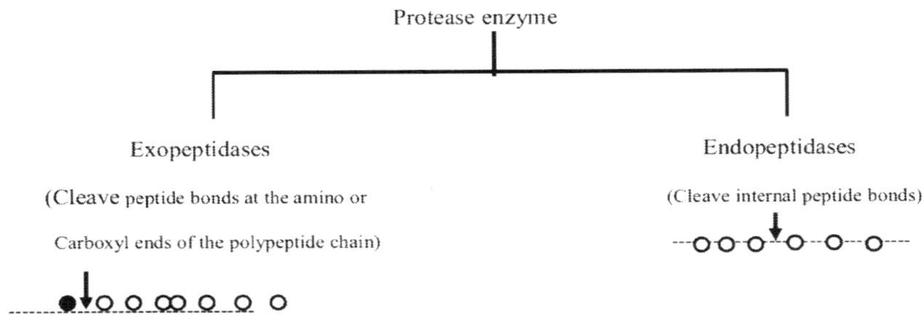

Figure 6.4. The mechanism of protease enzymes action. The proteases cleave proteins by a hydrolysis reaction-the addition of a molecule of water to a peptide bond.

Source: https://www.researchgate.net/figure/The-general-mechanism-of-protease-enzymes-action-The-Proteases-cleave-proteins-by-a_fig6_301308767.

Plant-based protease enzymes such as ficin, papain, and bromelain, have widespread applications in food industries including coagulation of milk, tenderization of meat, brewing, and as a digestive aid. Moreover, they also find the applications involving improvement in the nutritional value, flavor, solubility, digestibility of food proteins, and modifications in their functional features comprising emulsification and coagulation (Aruna et al., 2014). Protease enzymes are also extensively employed in baking industries to produce baked foods, crackers, bread, and waffles. They are utilized for decreasing dough consistency and uniformity, reducing the mixing time, regulating the gluten strength in bread, and for improvement in the flavor as well as texture. The acid protease associated with Aspergillus usamii is being efficaciously used to improve the wheat gluten functional properties (Deng et al., 2016). Adding protease enzyme could result in the discharge of an adequate quantity of peptides and amino acid levels in malt to acquire suitable fermentation. Acidic fungal protease enzymes are utilized for improved beer fermentation because of their proficiency even at reduced pH values by keeping the balance of the amino acid profile of the beer. An additional foremost application of protease enzymes is related to the dairy industry. Naturally ensuing protease enzymes are contributing considerably to enhance the flavor of cheese. These enzymes are employed for accelerating cheese ripening, modification of the existing functional properties as well as reducing the allergenic characteristics of milk-based products. In cheese production, protease enzymes have also been utilized for hydrolyzing the particular peptide bond to produce macropeptides and paracasein (Raveendran et al., 2018).

6.5. LACTASE (Β-GALACTOSIDASE) ENZYME

In the food industry, hydrolysis of lactose is a major biotechnological process. The hydrolysis of lactose can be catalyzed by the enzyme β-galactosidase. It is a unit of the family of hydrolases. Numerous biological systems, including microorganisms, plants, and animals are the source of β-Galactosidase. Due to higher yield, the production of the β-galactosidase enzyme from microorganisms such as bacteria, fungi, and yeast is a preferred choice, and hence it costs relatively lesser (Ladero et al., 2000). The final application of the enzyme β-galactosidase determines the selection of source, e.g., β-galactosidase from yeasts with optimum pH of 6.5–7.0 is commonly applied for the hydrolysis of lactose in whey milk. Fungal β-galactosidase with pH optima of 3.0–5.0 is appropriate in the case of hydrolysis of acidic whey. The final application of the enzyme or industry hence decides the selection of β-galactosidase. β-Galactosidase obtained from yeast Kluyveromyces lactis entails ions such as Mn^{2+} or Na^{+}, while Kluyveromyces fragilis needs Mg^{2+}, Mn^{2+} or K^{+} (Figure 6.5) (Jurado et al., 2002).

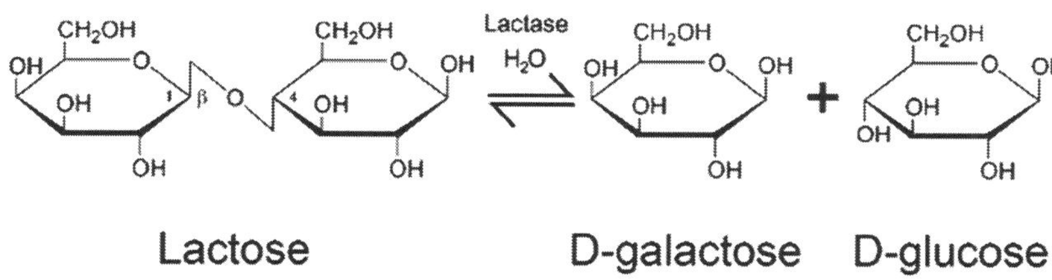

Figure 6.5. The enzymatic action of lactase hydrolyzing lactose to produce beta-D-galactose and beta-D-glucose.

Source: https://www.researchgate.net/figure/Enzymatic-action-of-lactase-Hydrolysis-of-lactose-by-lactase-to-produce-beta-D-galactose_fig2_326866705.

Two foremost types of β-galactosidases are of key significance in industrial applications. One is cold-active and the other is thermostable β-galactosidase. β-galactosidase is obtained through microorganisms with

GRAS status for use in milk and dairy products on a commercial scale. To reduce lactose intolerance in people, lactase is utilized with milk and milk-based products. Hydrolysis of lactose with lactase results in enhancement of the capability and creaminess of ice cream considerably. Another advantage of the hydrolysis of lactose into monomers is the reduced necessity of sweeteners by enhancing the sweetness of the products (Shah et al., 1993). The lactose hydrolysis in whey is another prime application of lactase. Minerals, lactose, and proteins are the major constituents of whey which is a byproduct during the production of cheese. The association of lactose with high biological oxygen demand (BOD) and chemical oxygen demand (COD) results in serious environmental problems related to the dairy industry (Johansen et al., 2002). The formation of galactooligosaccharides (GOS) from hydrolysis of lactose because of transglycosylation activity of β-galactosidase defines its further application. The GOS can be applied as a prebiotic food element.

6.6. LIPASES ENZYME

The enzymes catalyzing the hydrolysis of long-chain triglycerides are called lipases enzymes. They exist in the stomach as well as pancreas of humans and other animals for the digestion of fats and lipids. Yeast, bacteria, and fungi produce microbial lipases. 90% of the global lipase market covers microbial enzymes. These enzymes have widespread applications in various industries like food, detergents, animal feed, and biofuel industries. Other industries like textile, paper, and leather processing have also their vast applications (Gerits et al., 2014). In the food and beverage industries, lipases have important applications in fruit juice, dairy, wine, baking, and beer production. The market portion of lipase is less than 10% of the global industrial enzyme market even though it has numerous applications in different industries.

In dairy products, commercial lipases are mostly utilized for flavor improvement and processing of other foods having fat (Sharma et al., 2001). These enzymes could enhance the inherent flavor of cheese by performing activity on dairy fats to release free fatty acids during hydrolysis. From diverse sources, various varieties of cheese are produced through lipases, e.g., Camembert cheese using lipase from Penicillium camemberti, Romano cheese utilizing kid/lamb pre-gastric lipase, and cheddar cheese using Aspergillus niger or A. oryzae. The texture and softness of cheese can be improved by the catalysis of lipase. In butter and margarine, lipases are also utilized for the development of flavors and also to extend the shelf life of different baking products (Nema et al., 2019). The aroma could be modified by the use of lipase in alcoholic beverages like wine. They are employed to enhance the quality of required cocoa butter having a melting point of 37°C because of palmitic and stearic acids. Unilever has filed a patent suggesting the use of immobilized Rhizopus miehei lipase, that can substitute palmitic with stearic acid in order to provide required stearic-oleic-stearic triglyceride (Sharma et al., 2001). Functionalized phenols are esterified for producing lipophilic antioxidants which can be used in sunflower oil-consuming immobilized lipase from Helvina lanuginosa, Candida antarctica (CALB), Pseudomonas sp., Candida cylindracea Ay30 and Geotrichum candidum. In the food industry, lipases are used as a biosensor. Immobilized lipase has been effectively utilized for determining organophosphorus pesticides with a surface acoustic wave impedance sensor through lipase hydrolysis. In combination with glucose oxidase (GO), it can also be useful for finding triglycerides and blood cholesterol by restraining lipase onto pH/oxygen electrodes (Hasan et al., 2006). Lipase from Candida rugosa has numerous usages that could not be met by chemical synthesis. This lipase enzyme has applications in the production of single-cell protein, ice cream, carbohydrate esters, and derivatives of amino acids. Lipase can also be employed in the synthesis of different waste streams which are extracted by various food industries (Figure 6.6).

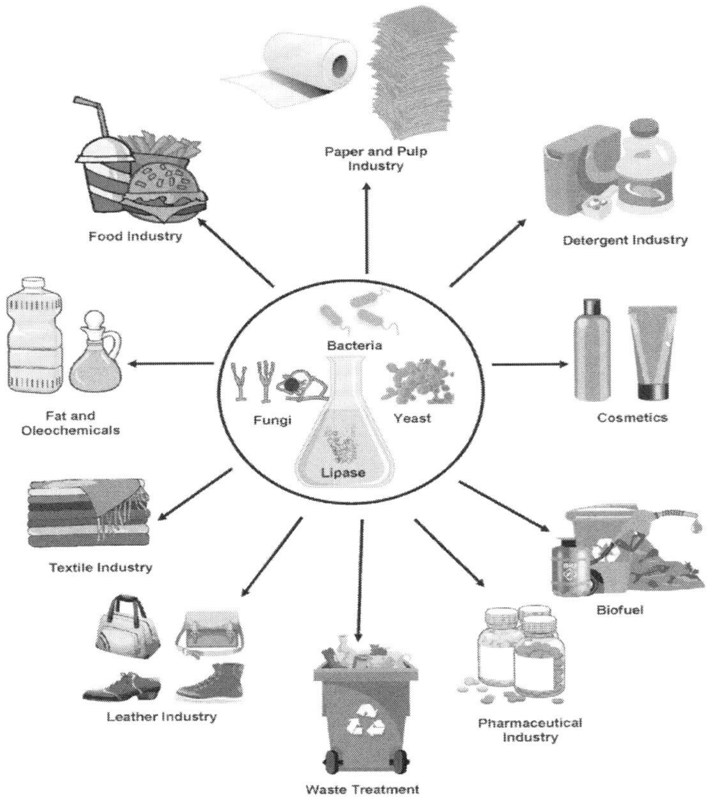

Figure 6.6. The microbial lipase and its use.

Source: https://link.springer.com/chapter/10.1007/978-981-15-1710-5_8.

6.7. PHOSPHOLIPASE ENZYME

Phospholipase enzymes break down phospholipids into fatty acids and further lipophilic substances, selectively. Based on their activity, it can be categorized in four key classes (A, B, C, and D) (Borrelli and Trono, 2015). The action of phospholipase A1 (PLA1), phospholipase A2 (PLA2), and phospholipase B on the carboxylic ester bonds of phospholipids, thus displace and replace the acyl group chain by different chemical reactions comprising hydrolysis, esterification, as well as transesterification (Figure 6.7).

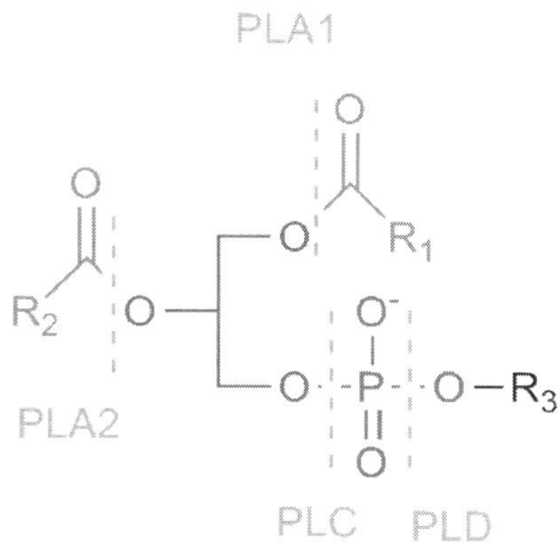

Figure 6.7. Phospholipase cleavage sites. An enzyme displaying both PLA1 and PLA2 activities is known as a phospholipase B.

Source: https://en.wikipedia.org/wiki/Phospholipase.

Phospholipases C (PLC) and D (PLD), modifying polar head groups, are also called phosphodiesterases recognizing the phosphodiester linkage. In the food industry, phospholipase enzymes are extensively utilized, especially in producing oils, dairy products, and in the manufacturing of numerous bakery products. They have also been used in the degumming of different vegetable oils, manufacturing of cheese as well as bread manufacturing (Maria et al., 2007). A commercially available phospholipase enzyme from Fusarium oxysporum having both phospholipase and lipase activities has been introduced by Novozymes A/S (Denmark) for baking applications with the trade name of LipopanF®. Zhao et al. (2010) studied PLA2, which was commercialized by DSM Food Specialties (The Netherlands) having trade name Maxapal®. He further stated that the egg yolk treatment with PLA may enhance the dough stability and interaction of starch and gluten. LysoMax® product (DSM food specialties) is another commercial phospholipase in the baking industry made by bacterial strain, explicitly act on lecithin. To improve the fat stability or increase the cheese quantity, milk, ice cream, and butter, phospholipases are useful in the synthesis of various dairy products (Maria et al., 2007). Vital uses of lipases comprise improvement in the cheese flavor, and production of lipolyzed milk fat for usage in butter as flavor, etc.

6.8. ESTERASE ENZYME

Esterase enzymes can assist the parting of esters into acid and alcohol in an aqueous solution. Being different from lipases, esterases hydrolyze short-chains instead of long-chain acylglycerols. In food industries and alcoholic beverages, esterases play an important role, where they are widely utilized to modify oil and fat in different fruit juices and synthesizing fragrances and flavors. Feruloyl esterases is a vital collection of enzymes belonging to the esterase family, that breakdown the ester bonds of ferulic acid with various polysaccharides in the plant cell walls. They are unavoidable for waste management as feruloyl esterases hydrolyze lignocellulosic biomass (Figure 6.8) (Faulds, 2010).

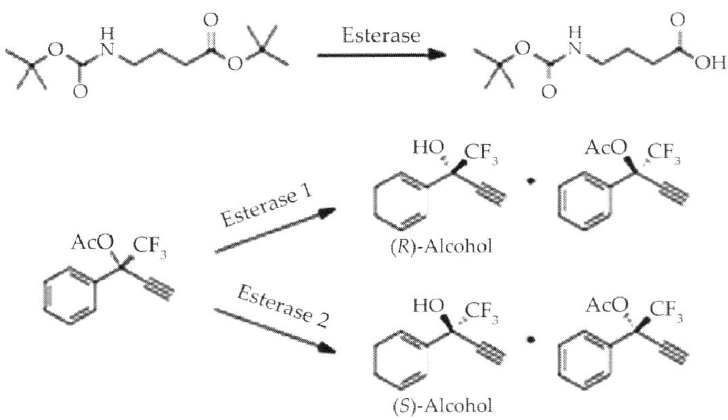

Figure 6.8. Example reactions for regio- and stereoselectivity of esterases.

Source: http://www.enzymicals.com/carboxylesterasen_en.html.

Cheng et al. (2012) vetted activity of feruloyl esterase in a metagenomic library which was acquired from the microbial population of a cow rumen and recognized a protease-resistant feruloyl esterase that could extract ferulic acid from wheat straws. Due to high pH, protease resistance as well as good thermal stability, this specific esterase has important commercial applications.

The fruity flavors are due to various methyl esters or ethyl esters of short-chain fatty acids in cheese manufacturing. Bacterial based production of thioesters and that of ethyl esters has also been studied by some researchers. Alvarez-Macarie and Baratti (2000) studied the making of an innovative thermostable esterase acquired from the highly thermotolerant Bacillus licheniformis heterologously described in E. coli for synthesizing short-chain flavor esters. Moreover, feruloyl esterase remains an important enzyme for biosynthesis of ferulic acid, that is used as a precursor for preparation of vanillin, which is an aroma compound utilized in various foods as well as beverages. Numerous researchers have studied the microbial synthesis of feruloyl esterase (Guglielmettie et al., 2008).

6.9. LIPOXYGENASE ENZYME

Lipoxygenases (LOX) are used for the deoxygenation of polyunsaturated fatty acids in lipids comprising a cis-1,4-pentadiene. These enzymes have a single polypeptide chain combined into domains of N-terminal and catalytic β-barrel. LOX are non-haem iron-containing enzymes. Catalytic reaction of LOX yields diverse precursors used to synthesize various volatile and aroma-generating chemical materials in plants. Enzymes of LOX are utilized in the generation of aroma in different food industries besides bread making. Mostly investigated lipoxygenase enzyme is the soya bean LOX. Bacterial LOXs have various specificities to fatty acids. LOX enzymes from Nostoc sp. oxygenate at a particular site in linoleic acid, nevertheless, the LOX enzyme from Anabaena sp. displays an adjustable specificity (Figure 6.9) (Gao et al., 2010).

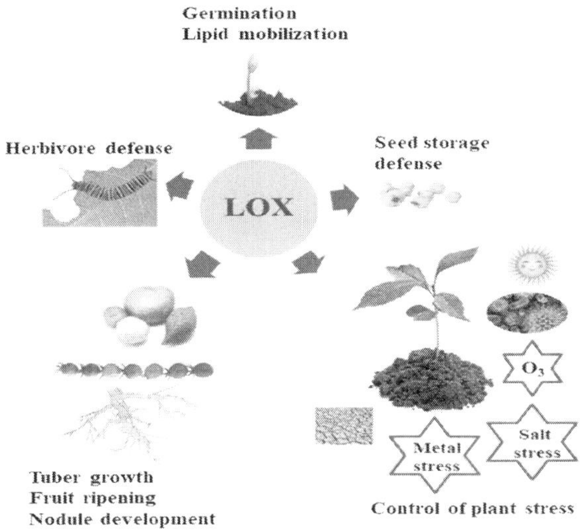

Figure 6.9. The use of lipoxygenases in the physiology of plants.

Source: https://link.springer.com/article/10.1007/s12374-020-09241-x.

The key application of LOX enzymes in the dough is based on its property to decolorize the flour pigment carotenoid, through co-oxidation of the dye with fatty acid. LOX have also been applied to increase tolerance in mixing and handling characteristics of dough. This impact originates because of thiol group oxidation in gluten, leading to rearrangement of various disulfide bonds, tyrosine cross-linking, and further solidification of the gluten. It is also helpful in improving the dough rheology. Lately, researchers decontaminated lipoxygenase from Lasiodiplodia theobromae by various chromatography methods and analyzed the full characteristics of the enzyme. L. theobromae has been studied to encompass two kinds of LOX of molecular mass of 93 and 45 kDa.

6.10. CELLULASE ENZYME

Cellulases are the type of enzymes that convert polymeric cellulose as well as β-1,4 linkages into glucose molecular units. There are three important types of cellulases: endo-(1,4)-β-d-glucanase, exo-(1,4)-β-d-glucanase and β-glucosidases (Patel et al., 2015). Cellulases with catalytic behavior are mostly glycoside hydrolases (GH) family and are categorized into various groups on the basis of amino acid arrangements variability and its structural three-dimensional characteristics. The family of GH enzymes mostly utilizes an acid-base catalysis mechanism for cleaving glycoside bonds in cellulose. The process of catalysis is mostly dependent on two key residues (a nucleophile and a proton donor) of the enzyme in the active site (Sinnott, 1990). Hydrolysis takes place due to a retention or inversion mechanisms that depend on the three-dimensional position of the catalytic residues present in the enzyme. Non-reducing and reducing ends of the cellulose polymer are exposed by the endoglucanases cleave β-1,4-bonds present in the amorphous region of cellulose. Different sources of endoglucanases are associated with various families of glycoside hydrolase, amongst which the important are 5–9, 12, 44, 45, 48, 51, and 74. Most fungal endoglucanases have a catalytic module with a carbohydrate-binding module (CBM), although the catalytic module deprived of CBM has also been conveyed from fungal species. Many catalytic modules and CBMs exist as bacterial endoglucanases. The catalytic module of the majority of the endoglucanases has a cleft/grove-shaped active site (Figure 6.10) (Payne et al., 2015).

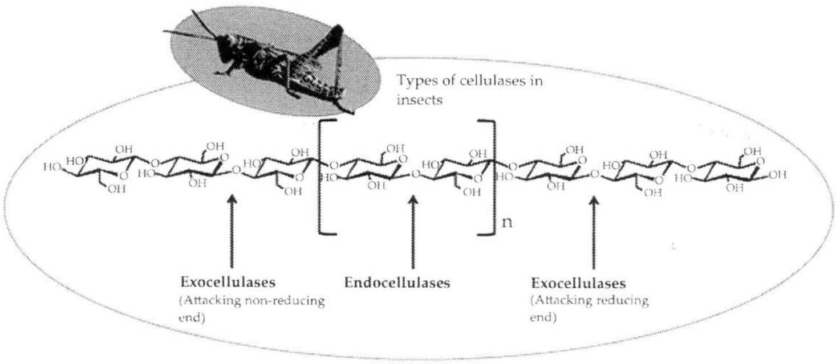

Figure 6.10. Cellulase enzymes extracted from insects.

Source: https://link.springer.com/chapter/10.1007/10_2013_206.

Exoglucanases or cellobiohydrolases (CBHs) react with the reducing ends or non-reducing ends of available cellulose polymer to produce cellobiose. Diversity in fungal and bacterial CBHs in catalytic module belongs to glycoside hydrolase families 5, 6, 7, 9, 48, and 74. Mostly the CBHs are witnessed in a tunnel-shaped catalytic module. The cleaving of the non-reducing terminal β-d-glucosyl residues and removal of β-d-glucose is achieved with β-Glucosidases catalysis in the final step in cellulose breakdown (Kigel et al., 1995). From different β-glucosidases, the catalytic modules related to glycoside hydrolase families 1, 3, and 9 have been conveyed. The regulation of cellulolytic machinery of microbes is primarily through feedback inhibition of β-glucosidases by their reaction product glucose. The absence of CBM in their structure is its main differentiation from CBHs. A pocket-shaped active site region of β-glucosidases aids to attach the glucose molecule to the non-reducing end and release the glucose element from cellodextrins or cellobioses (Payne et al., 2015). During the growth of cellulosic materials, a large variety of microorganisms have been reported to produce celluloses. There is a diversity in the habitat of microorganisms like bacteria and fungi, which are the main microbial sources for industrial synthesis.

Aerobic fungi and aerobic bacteria have the comparable mechanism of cellulose degradation. Cellulosomes are present on the cell surface of anaerobic bacteria and behave via an altered system. The potential usage of cellulases from fungi (Aspergillus and Trichoderma) and bacteria (Bacillus and Paenibacillus) is in food production. Besides, they also have an important use in different industries, including textile, detergent, food, and paper industries (Mekala et al., 2008). In the juice industries, cellulases are used in mixtures with other macerating enzymes for enhanced performance of process to acquire high yield, by improving the extraction systems, clarification, and stabilization of juices. The viscosity of nectar and puree can also be reduced from fruits such as papaya, apricot, pear, plum, peach, mango, and are utilized to extract flavonoids from various flowers and seeds. Cellulase-mediated extraction is preferred over conventional techniques because of greater yield, minimum heat damage as well as less processing time. The extraction of phenolic compounds from grape pomace is usually done from cellulases (Kabir et al., 2015). β-Glucosidases as blend with pectinase change the structure, flavor, and fragrance of the utilized fruits and vegetables. They have also been reported for the reduction of bitterness of citrus fruit and improvement in fragrance and taste. Cellulases have an application with other enzymes for proficient olive oil extraction. Cellulose enzymes have been utilized mixture with other enzymes to raise the yield and quality of produced wine. The main benefits of utilizing such enzymes are better maceration, color development, clarification, wine stability, and quality. A study carried out by Oksanen et al. (1985) demonstrated that cellulases can efficiently decrease wort viscosity. The aroma of wines could be enhanced by β-glucosidases via alterations of glycosylated precursors.

6.11. XYLANASE ENZYME

Xylanases are formed by microorganisms to cleave xylans which is a key component of hemicellulose. Synergistic action of three main enzymes, i.e., endoxylanases, exoxylanases, and β-xylosidases is essential for the breakdown of the xylan backbone in hemicellulose. β-1,4 bonds of the xylan backbone are cleaved by endoxylanases. Exoxylanase enzymes hydrolyze β-1,4 bonds of xylan from the non-reducing ends and produce xylooligosaccharides. β-xylosidases cleave the xylobiose and xylooligosaccharides to discharge xylose (Sukumaran, 2009). The catalytic module performs the major functions of xylanases and few classes have a further CBM to bond with the substrates. GHs and carbohydrate esterases (CEs) are the two major catalytic modules of hemicellulases. Xylan backbone is hydrolyzed by the endoxylanase enzymes and has catalytic cores which belong to GH families 8, 10, 11, 30, and 43 having utmost common ones being GH 10 and 11. GH10 is more active on substituted xylan and also these are different in their substrate specifications. They could comprise CBMs similar to cellulases. Exoxylanases arbitrarily cleave the xylan backbone from inside with the release of long-chain xylooligomers on which the β-xylosidases actions take place. The catalytic module of such enzymes are associated with the GH families 3, 30, 39, 43, 52, and 54. These two enzymes together are called xylanases. The action of β-Xylosidase or xylan-1,4-β-xylosidase on the xylooligosaccharides and xylobiose produces xyloses (Figure 6.11) (Sanghi et al., 2010).

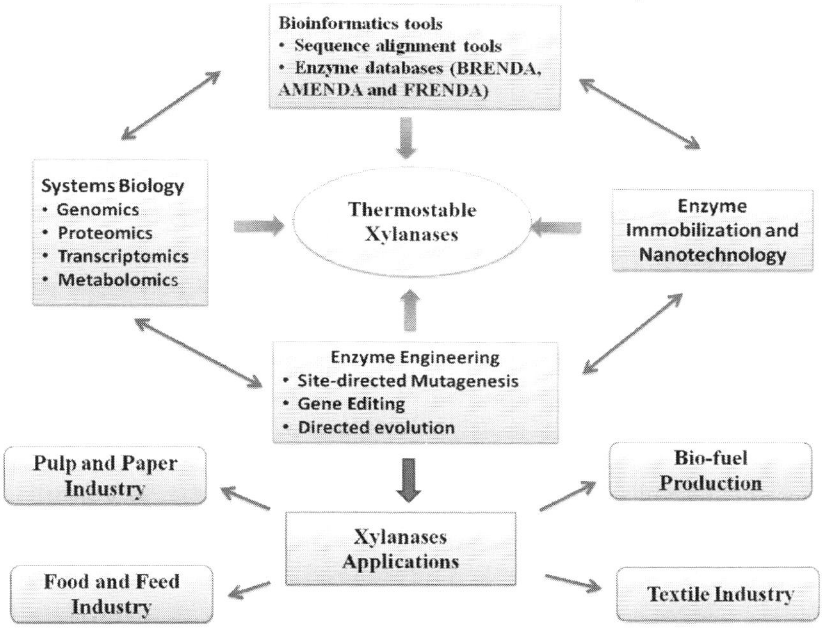

Figure 6.11. The modern techniques to increase the thermostability of xylanases in industrial applications.

Source: https://www.researchgate.net/figure/Modern-techniques-for-increasing-the-thermostability-of-xylanases-in-industrial_fig1_322781886.

The microbes including actinomycetes, bacteria, and fungi produce xylanases. Species of actinomycete and bacteria which yield producing xylanase enzymes comprise Streptomyces sp., Bacillus sp., and Pseudomonas sp. Xylanase enzymes produced by species of bacteria and actinomycetes generally act in a wide range of pH (5.0–9.0) at activity temperature of 35°C to 60°C. The foremost sources of xylanase are fungi because of high content and extracellular enzyme released by them (Mandal, 2015). Important fungi species yielding xylanase are Fusarium sp., Aspergillus sp., and Penicillium sp. The activity of bacteria or yeast is lower than fungal xylanases.

Enzymes that hydrolyze carbohydrates are mostly utilized in bread-making industries. The enzymatic hydrolysis of non-starch polysaccharides improves the rheological properties of dough (Polizeli et al., 2005). Bread making industry widely utilizes a mixture of xylanase enzymes with different enzymes due to the prospective efficiency of xylanase enzymes in the production of bread. It improves the volume and the quality of produced bread. Xylanase enzymes break down hemicellulose present in wheat flour, improving water binding in the dough. Consequently, the dough develops soft and crumbs formation delays which allow the dough to rise. Xylanase enzyme is also utilized to develop quality, flavor, and deliciousness in biscuits. These enzymes also perform a significant role in the production of juice by enhancing extraction, clarification of juice, as well as stabilization. Xylanases lead to better yield in combination with other enzymes of juice and escalate retrieval of fragrances, vitamins, mineral salts, essential oils, pigments, etc. (Saadoun et al., 2013). Furthermore, xylanase enzymes are utilized for hydrolyzing the cellular wall of barley in producing beer. Hydrolysis leads to the discharge of arabinoxylans and lower oligosaccharides, declining the indistinct look and viscosity of the beer.

6.12. PECTINASE ENZYME

Glycosidic bonds existing in the pectic polymers are hydrolyzed by the catalytic activity of pectinase enzymes. Pectic materials found in apple, lemon pulp, pineapple, orange, tomato, orange peel, and additional citrus fruits serve as a natural substrate for pectinase enzyme. Pectinases are divided functionally as pectin lyase, polygalacturonases (hydrolyzing glycosidic α-(1-4) bonds), pectin esterases (removing acetyl groups and methoxyl groups from pectin), and pectate lyase. Pectinases are acquired naturally and from microbes with efforts made to enhance their thermal stability besides yield. Pectinases could also be used either on flat or hairy areas of pectin (Tapre and Jain, 2014). Based on the value of pH, there are alkaline and acidic pectinases which are also convened in endopectinases-when the enzyme cleaves unevenly, and exopectinases-when the terminal ends are targeted (Figure 6.12).

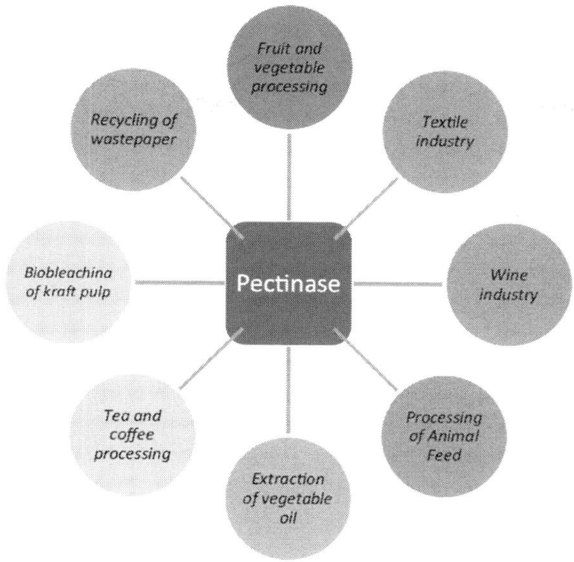

Figure 6.12. Applications of pectinases.

Source: https://link.springer.com/article/10.1007/s13205-016-0371-4.

Pectinases find diverse industrial applications comprising the food industries, remediation, paper bleaching, etc. Juices with mixed pectinase possess a flawless appearance and filterability compared to the enzyme-depleted counterparts. Pectinase enzymes also enhance the color and flavor of drinks besides decreasing the turbidity and haze age group of naturally derived fruit juices such as apple and banana (Muller, 1928). Adding gelatin and pectin to juices raises the viscosity as well as turbidity of produced juices, and exclusion of the haze of juice production costs high. To achieve better outcomes, utilization of biogenic enzymes like pectinases in juices would result in nine times better outcome than mechanical softening.

6.13. GLUCOSE OXIDASE (GO) ENZYME

Glucose oxidase (GO) is a group of enzyme families termed oxidoreductases. In 1928, Müller discovered GO which is flavoprotein. He explained that under dissolved oxygen the enzyme could transform glucose into gluconic acid. Later, β-d-glucose is oxidized to gluconolactone and oxygen is reduced to hydrogen peroxide in this reaction. Gluconolactone is later instinctively hydrolyzed to gluconic acid (Figure 6.13) (Muller, 1928).

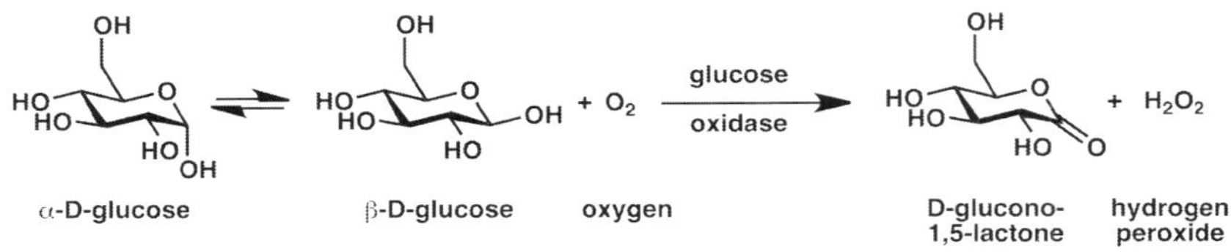

Figure 6.13. The reaction catalyzed by glucose oxidase.

Source: https://en.wikipedia.org/wiki/Glucose_oxidase.

The homodimeric enzyme comprises two identical polypeptide chain subunits (80 kDa). Disulfide bonds covalently link the subunits and one flavin adenine dinucleotide (FAD) molecule non-covalently bound to the active site area of every subunit. Different organisms have been reported for the production of GO, and it has been initially revealed in Penicillium glaucum and Aspergillus niger. Aspergillus niger species has been extensively utilized for the production of GO, and a higher amount of GO can be produced from its strains (Mikhaĭlova et al., 2000). Extracellular GO is created from a widely used fungus named Penicillium adametzii. GO production has also been reported from different bacterial species. Though numerous bacterial and fungal species are conveyed to yield this enzyme, whereas fungi have opted for commercial yield of GO (Giardina et al., 2010). Pharmaceutical and food industries widely utilize GO in addition to its use in biofuel cells. Its increased usage in baking industries is due to its oxidizing properties, which produce stronger dough. In applications of food industries, it improves the fragrance, stability, and flavor, of food products by removal of glucose in addition to oxygen from egg white and diabetic drinks. The improvement in color, texture, flavor, and shelf life of food items and prevention of rotting can be possible by GO. In food packing, GO is utilized for enhancement in shelf life by the removal of oxygen (Hanft and Koehler, 2006).

6.14. LACCASE ENZYME

Laccases, a group of oxidases signifies the largest subgroup of multicopper enzymes. They are usually blue oxidases which are utilized for their potential oxidation of phenolic compounds and hence applied in various industrial sectors. They act as potent biocatalyst for biobleaching of paper pulp, wine stabilization, bioremediation, chemical synthesis, biosensing, and textile finishing. They have diverse qualifications for broad variety of oxidizable substrates that are further dependent on the type of microbial sources that produce the enzyme (Morozova et al., 2007). The oxidation of a wide variety of compounds has been catalyzed by laccases like aromatic amines, phenolics, and ascorbate. The combination of these enzymes reduces substrate comprising four oxidized electrons and four reduced electrons to cleave dioxygen bond in the existence of laccases having four copper atoms. The catalytic activity mechanism of laccase has been explained by researchers (Gochev and Krastanov, 2007).

6.15. CATALASE ENZYME

A tetrameric protein, catalase exists in aerobic organisms. It is helpful in the decomposition of hydrogen peroxide. Microbial sources named Micrococcus luteus, Aspergillus niger, and bovine liver are used for the production of this enzyme. Microorganisms are generally favored as sources of production of various enzymes because of proficiencies, including easy handling, fast growth, and genetic tuning to acquire the required product (Haruyuki and Tadayuki, 2003). Bacteroides fragilis (anaerobic) unveiled enhanced catalase levels in media with haem. Singh et al. (2007) separated the catalase enzyme from Enterococcus faecalis, entirely dependent on a haem source whose absence could not produce the porphyrin group. A facultative anaerobic catalase-producing Bacillus maroccanus immune to hydrogen peroxides was separated from textile wastes (Figure 6.14) (Singh et al., 2007).

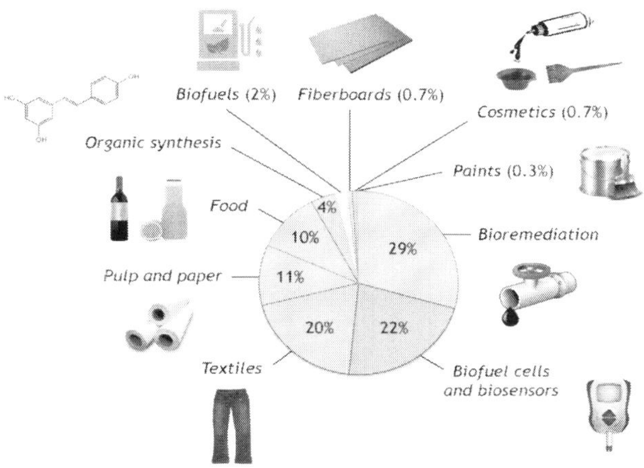

Figure 6.14. The breakdown of the biotechnological applications of laccases.

Source: https://sfamjournals.onlinelibrary.wiley.com/doi/full/10.1111/1751-7915.12422.

In Los Banos and Calamba, Laguna, Philippines, a potent catalase-producing bacterium Pyrobaculum calidifontis was separated from hot springs. A thermo-alkaliphilic catalase-positive strain of Bacillus halodurans LBK 261 was separated from alkaline hot-spring waters of Kenya. Isolation was done for halo (alkali) tolerant catalase-producing Oceanobacillus oncorhynchi ssp. incaldaniensis from an algal mat having the capability to produce catalase at a broad range of pH 6.0–9.5 with the salinity of 5–20% (Wang et al., 2008). From seawater, a catalase-positive psychrophile Bacillus N2a was extracted (Nakayama et al., 2008). From industrial effluents from beverage industries, catalase-positive bacteria called Rhizobium radiobacter has been isolated, Comamonas testosteroni and C. terrigena were isolated from wastes slurry supplemented with crude oil together with heavy metals and Serratia SYBC08 from hydrogen peroxide sludge. A psychrotolerant bacterium such as Psychrobacter piscatorii T-3 has been separated from bleach-rich overspill, possesses elevated activity of catalase (Kimoto et al., 2012). Similarly, fungi, and yeast have the capability of producing catalase. In isolates of Aspergillus niger, the utmost level of catalase activity of 400 mg/g has been detected. A catalase-positive entomopathogenic fungus Metarhizium anisopliae strain Ma10 (CNRCB MaPL10) has been liberated from Geraeus senilis.

Catalase is also utilized for the removal of surplus hydrogen peroxide from fabrics in fabric industries. In food processing industries, this enzyme is typically utilized together with other enzymes. For the preservation of food items, catalase is frequently employed with GO enzymes. Ough (1975) has utilized a GO/catalase cocktail for eliminating the oxygen from wine before bottling and assessed the development of acetaldehydes. The outcomes revealed the stability of color and amount of acetaldehyde with proper treatment with enzymes (Röcker et al., 2016). Catalase is used for the removal of peroxide from milk in the milk processing industries to eradicate glucose from egg white in the baking industry and food wrappers to resist oxidation and to regulate the perishability of food items. Cheese production involves limited usage of this enzyme.

6.16. PEROXIDASE PROTEINS

Peroxidases are oxidoreductase proteins containing iron(III) protoporphyrin IX as the prosthetic group. The reduction of peroxides as well as oxidation of a broad variety of various inorganic and organic compounds is catalyzed by these proteins. They possess molecular mass ranging from 30,000 to 150,000 Da and comprising a group of distinctive enzymes NADH peroxidase, like iodide peroxidase, glutathione peroxidase, and a group of other general enzymes (Hamid, 2009). Plants, microorganisms, and animals contain peroxidases. Lignification processes are governed by peroxidases in various plants besides defense mechanisms against injured or contagious tissues.

Phanerochaete chrysosporium is considered as the preeminent peroxidase-secreting microorganism compared to all other microorganisms. The challenges of post-translational modification of proteins limit the large-scale industrial applications of fungal peroxidases. Nevertheless, bacterial peroxidases are relatively easy to produce with improved stability and activity that is acceptable for industrial applications. For dye decolorization, enzymes are utilized with bacterial laccases (Colpa et al., 2014). Peroxidase activities are conveyed in bacterial taxa like Proteobacteria, Actinobacteria, Acidobacteria, and Firmicutes. Furthermore, actinomycetes, a soil bacteria, can propagate like fungi and have alike ecological niches and could yield peroxidases for the degradation of lignin. Streptomyces viridosporus T7A was the first to produce extracellular lignin peroxidase (Figure 6.15) (Ramachandra et al., 1988).

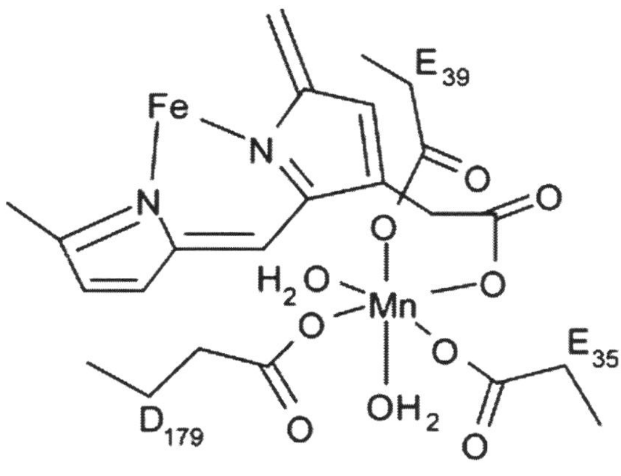

Figure 6.15. The typical structure of peroxidase.

Source: https://www.sciencedirect.com/topics/medicine-and-dentistry/peroxidase.

Peroxidase catalysis is for a large variety of substrates that use hydrogen peroxide or other peroxides. In the food industry, this enzyme is utilized to produce color, texture, and flavor as well as to improve the nutritional quality of food. Further applications involve polymer synthesis, biosensors, and environmental waste management (Singh et al., 2020). It has also been used for the treatment of phenolic discharges from the industry. In food industries, thermal inactivation of peroxidases is employed to evaluate the effectiveness of blanching treatment, which improves the shelf life of various foods (Regalado et al., 2004). The unwanted browning of fruits and off-flavors of vegetables is due to the adverse effects of peroxidases.

6.17. A-ACETOLACTATE DECARBOXYLASE ENZYME

α-Acetolactate decarboxylase is used for enhancing the fast maturation of beer. These enzymes could be acquired naturally from microbes such as Brevibacillus brevis or recombinant Saccharomyces cerevisiae. This enzyme catalyzes the conversion of acetolactate to acetoin through a two-step reaction. The first step involves direct decarboxylation of the substrate to an enol derivative, and the second step entails further protonating to the final product (Zhao et al., 2016).

Enzyme-based elimination of α-acetolactate and α-aceto-α-hydroxybutyrate helps in overwhelming the rate-limiting step of beer maturation. Beer maturation without the usage of enzymes takes 2 to 12 weeks, depending on the source of the enzyme, α-acetolactate decarboxylase matures the beer within 24 hours. Furthermore, the off-taste because of the existence of diacetyl in beer is invalidated by these enzymes. Both free and encapsulated forms of this enzyme performed fine in the process as conveyed in previous studies, thus promoting the utilization of immobilized enzymes at abridged costs. For enhanced thermal stability, catalytic efficiency, and reusability, the use of innovative inorganic nanoflowers or alginate microbeads restrained with α-acetolactate decarboxylase are auspicious approaches (Figure 6.16) (Zhao et al., 2017).

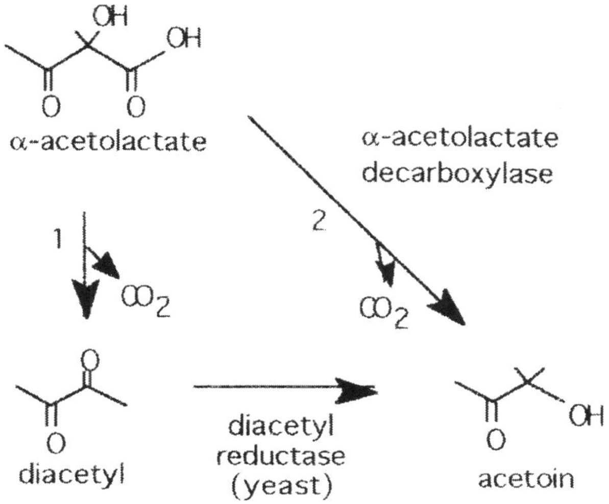

Figure 6.16. The reaction catalyzed by-acetolactate decarboxylase. 1: oxidative decarboxylation; 2: enzymatic nonoxidative decarboxylation.

6.18. ASPARAGINASE ENZYME

Asparaginases possess a foremost class of nutraceutical, pharmaceutical, and industrially substantial enzymes among numerous microbially resulting enzymes that are extensively utilized by man. Asparaginase catalyzes the breakdown of the asparagine to resultant acid derivative aspartic acid and NH_3 and could be deliberated as an asparagine-depleting enzyme. To humans, asparagine is a dispensable amino acid, though it is a significant amino acid for cancerous cells. Hence, the development of cancerous cells is unfavorably affected by the depletion of asparagine, which provides the basis of these enzymes as an anticancerous agent (Krishnapura et al., 2016).

Different food processing techniques comprising boiling in oil and baking result in the transformation of asparagine to acrylamide, an identified carcinogen. Among several approaches endeavoring to overwhelm the acrylamide development, the reduction of asparagines through enzymatic treatment has proven to be more operative in decreasing the development of acrylamides from asparagines by at least 97% (Figure 6.17).

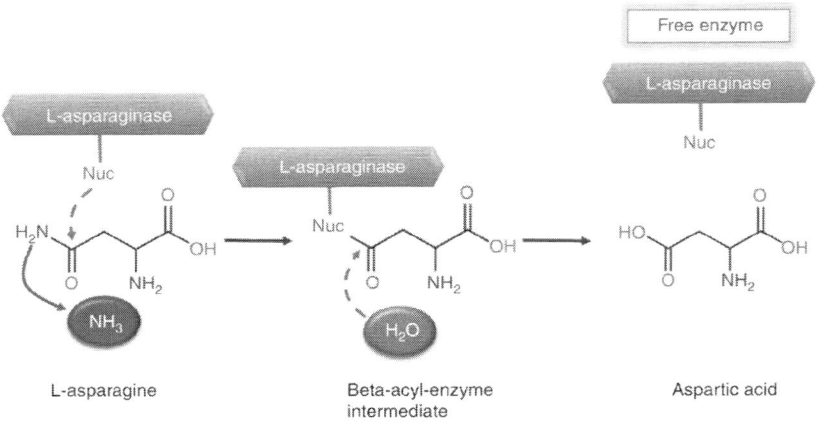

Figure 6.17. The general mechanism of L-asparaginase catalyzed reaction. Dashed arrow displays the nucleophilic attack.

Source: https://www.scielo.br/j/bjm/a/DPgVGPcH8sBfRsjhdnSW59y/?lang=en.

6.19. DEBITTERING ENZYME-NARINGINASE

Naringinase is generally used for the dissociation of naringin, the standard unpleasant flavanone glycoside existing in citrus fruits. Naringin is converted to a glycon naringenin and rhamnose due to α-rhamnosidase and β-glucosidase action. These are mostly formed by fungal isolates, viz. Aspergillus niger, Fusarium, Penicillium, Rhizopus, Eurotium, Circinella, and Trichoderma and bacteria such as Bacteroides distasonis, Bacillus sp., Burkholderia cenocepacia, Pseudomonas paucimobilis, Thermomicrobium roseum, etc. (Puri, 2012). Due to amplified yield, fungal sources of naringinase have more significant use compared to bacterial ones (Figure 6.18).

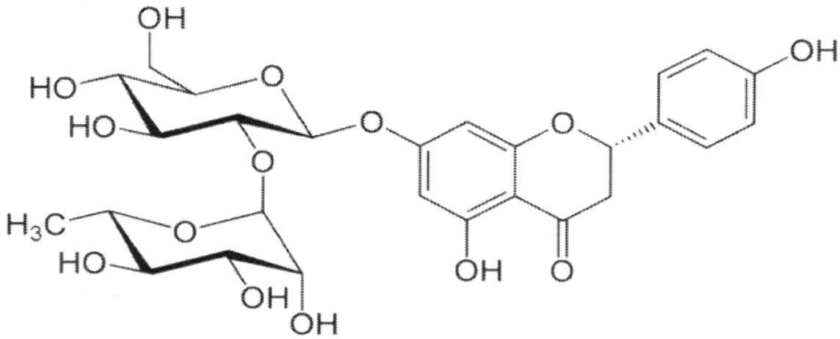

Figure 6.18. The typical structure of Naringin.

Source: https://en.wikipedia.org/wiki/Naringin.

Naringinase plays a vital role in the processing of food as a debittering enzyme added to fruit juices. To acquire improved results, both free and restrained forms of these enzymes are utilized. Immobilization of these enzymes could be obtained in a range of substrates like polyvinyl alcohol cryogel, cellulose triacetate nanofiber graphene, packaging films, etc. (Xiang et al., 2020). Different food flavorings comprising biopolymers and sweeteners could be produced by rhamnosidase or naringinase. However, additional usage of naringinase with β-glucosidase and arabinosidase assists in improving the fragrance of wine. The application of naringinase has also been found in kinnow peel waste management, prunin preparation, and tomato pulp preparation (Puri, 2012).

REFERENCES

1. Alvarez-Macarie, E., & Baratti, J., (2000). Short-chain flavor ester synthesis by a new esterase from Bacillus licheniformis. Journal of Molecular Catalysis B: Enzymatic, 10(4), 377–383.
2. Aruna, K., Shah, J., & Birmole, R., (2014). Production and partial characterization of alkaline protease from Bacillus tequilensis strains CSGAB0139 isolated from spoilt cottage cheese. Int. J. Appl. Biol. Pharm., 5, 201–221.
3. Blanco, C. A., Caballero, I., Barrios, R., & Rojas, A., (2014). Innovations in the brewing industry: Light beer. International Journal of Food Sciences and Nutrition, 65(6), 655–660.
4. Borrelli, G. M., & Trono, D., (2015). Recombinant lipases and phospholipases and their use as biocatalysts for industrial applications. International Journal of Molecular Sciences, 16(9), 20774–20840.
5. Cai, L., Nian, L., Cao, A., Zhang, Y., & Li, X., (2020). Effect of carboxymethyl chitosan magnetic nanoparticles plus herring antifreeze protein on conformation and oxidation of myofibrillar protein from red sea bream (Pagrosomus major) after freeze-thaw treatment. Food and Bioprocess Technology, 13(2), 355–366.
6. Camacho, N. A., & Aguilar, G., (2003). Production, purification, and characterization of a low-molecular-mass xylanase from Aspergillus sp. and its application in baking. Applied Biochemistry and Biotechnology, 104(3), 159–171.
7. Cheng, F., Sheng, J., Cai, T., Jin, J., Liu, W., Lin, Y., & Shen, L., (2012). A protease-insensitive feruloyl esterase from China Holstein cow rumen metagenomic library: Expression, characterization, and utilization in ferulic acid release from wheat straw. Journal of Agricultural and Food Chemistry, 60(10), 2546–2553.
8. Colpa, D. I., Fraaije, M. W., & van Bloois, E., (2014). DyP-type peroxidases: A promising and versatile class of enzymes. Journal of Industrial Microbiology and Biotechnology, 41(1), 1–7.
9. Coutinho, P. M., & Reilly, P. J., (1997). Glucoamylase structural, functional, and evolutionary relationships. Proteins: Structure, Function, and Bioinformatics, 29(3), 334–347.
10. Couto, S. R., & Sanromán, M. A., (2005). Application of solid-state fermentation to ligninolytic enzyme production. Biochemical Engineering Journal, 22(3), 211–219.

11. De Maria, L., Vind, J., Oxenbøll, K. M., Svendsen, A., & Patkar, S., (2007). Phospholipases and their industrial applications. Applied Microbiology and Biotechnology, 74(2), 290–300.
12. Deng, L., Wang, Z., Yang, S., Song, J., Que, F., Zhang, H., & Feng, F., (2016). Improvement of functional properties of wheat gluten using acid protease from Aspergillus usamii. PLoS One, 11(7), e0160101.
13. Faulds, C. B., (2010). What can feruloyl esterases do for us? Phytochemistry Reviews, 9(1), 121–132.
14. Gao, B., Boeglin, W. E., & Brash, A. R., (2010). Omega-3 fatty acids are oxygenated at the n-7 carbon by the lipoxygenase domain of a fusion protein in the cyanobacterium Acaryochloris marina. Biochimica et Biophysica Acta (BBA)-Molecular and Cell Biology of Lipids, 1801(1), 58–63.
15. Gerits, L. R., Pareyt, B., Decamps, K., & Delcour, J. A., (2014). Lipases and their functionality in the production of wheat-based food systems. Comprehensive Reviews in Food Science and Food Safety, 13(5), 978–989.
16. Giardina, P., Faraco, V., Pezzella, C., Piscitelli, A., Vanhulle, S., & Sannia, G., (2010). Laccases: A never-ending story. Cellular and Molecular Life Sciences, 67(3), 369–385.
17. Gochev, V. K., & Krastanov, A. I., (2007). Isolation of laccase producing Trichoderma spp. Bulgarian Journal of Agricultural Science, 13(2), 171.
18. Guglielmetti, S., De Noni, I., Caracciolo, F., Molinari, F., Parini, C., & Mora, D., (2008). Bacterial cinnamoyl esterase activity screening for the production of a novel functional food product. Applied and Environmental Microbiology, 74(4), 1284–1288.
19. Hamid, M., (2009). Khalil-Ur-Rehman. Food Chem., 115(4), 1177–1186.
20. Hanft, F., & Koehler, P., (2006). Studies on the effect of glucose oxidase in bread making. Journal of the Science of Food and Agriculture, 86(11), 1699–1704.
21. Haruyuki, A., & Tadayuki, I., (2003). High temperature environment adaptation strategy for hyperthermophilic bacteria. Biochemistry, 75 (7), 561–575.
22. Hasan, F., Shah, A. A., & Hameed, A., (2006). Industrial applications of microbial lipases. Enzyme and Microbial Technology, 39(2), 235–251.

23. Johansen, A. G., Vegarud, G. E., & Skeie, S., (2002). Seasonal and regional variation in the composition of whey from Norwegian Cheddar-type and Dutch-type cheeses. International Dairy Journal, 12(7), 621–629.
24. Jooyandeh, H., Amarjeet, K., & Minhas, K. S., (2009). Lipases in dairy industry: A review. Journal of Food Science and Technology (Mysore), 46(3), 181–189.
25. Jurado, E., Camacho, F., González-Tello, P., & Robles, A., (1996). Microcrystalline-cellulose hydrolysis with concentrated sulfuric acid. Journal of Chemical Technology and Biotechnology: International Research in Process, Environmental AND Clean Technology, 67(4), 350–356.
26. Jurado, E., Camacho, F., Luzon, G., & Vicaria, J. M., (2002). A new kinetic model proposed for enzymatic hydrolysis of lactose by a β-galactosidase from Kluyveromyces fragilis. Enzyme and Microbial Technology, 31(3), 300–309.
27. Kabir, F., Sultana, M. S., & Kurnianta, H., (2015). Polyphenolic contents and antioxidant activities of underutilized grape (Vitis vinifera L.) pomace extracts. Preventive Nutrition and Food Science, 20(3), 210.
28. Kigel, J., (1995). Biochemical and molecular characterization of a barley seed-glucosidase. Journal of Biological Chemistry, 270(26).
29. Kimoto, H., Yoshimune, K., Matsuyma, H., & Yumoto, I., (2012). Characterization of catalase from psychrotolerant Psychrobacter piscatorii T-3 exhibiting high catalase activity. International Journal of Molecular Sciences, 13(2), 1733–1746.
30. Kohlhepp, G., (2010). Analysis of the situation of ethanol and biodiesel production in Brazil. Advanced Studies.
31. Kona, R. P., Qureshi, N., & Pai, J. S., (2001). Production of glucose oxidase using Aspergillus niger and corn steep liquor. Bioresource Technology, 78(2), 123–126.
32. Krishnapura, P. R., Belur, P. D., & Subramanya, S., (2016). A critical review on properties and applications of microbial l-asparaginases. Critical Reviews in Microbiology, 42(5), 720–737.
33. Kumar, N. M., Shimray, C. A., Indrani, D., & Manonmani, H. K., (2014). Reduction of acrylamide formation in sweet bread with L-asparaginase treatment. Food and Bioprocess Technology, 7(3), 741–748.

34. Ladero, M., Santos, A., & Garcıa-Ochoa, F., (2000). Kinetic modeling of lactose hydrolysis with an immobilized β-galactosidase from Kluyveromyces fragilis. Enzyme and Microbial Technology, 27(8), 583–592.

35. Law, B. A., (2010). Enzymes in dairy product manufacture. Enzymes in Food Technology, 88–102.

36. Law, B., (2002). Enzymes in the manufacture of dairy products. Enzymes in Food Technology (Vol. 19, pp. 91–108). Sheffield Academic Press Ltd. Mansion House.

37. Magalhães, O., & Souza, M., (2010). Application of microbial amylase in industry: A review. Brazilian Journal of Microbiology, 41, 850–861.

38. Mandal, A., (2015). Review on microbial xylanases and their applications. Appl. Microbiol. Biotechnol., 42, 45–42.

39. Mekala, N. K., Singhania, R. R., Sukumaran, R. K., & Pandey, A., (2008). Cellulase production under solid-state fermentation by Trichoderma reesei RUT C30: Statistical optimization of process parameters. Applied Biochemistry and Biotechnology, 151(2), 122–131.

40. Mikhaĭlova, R. V., Shishko, Z. F., Yasenko, M. I., & Lobanok, A. G., (2000). Effect of culture conditions on extracellular glucose oxidase production by Penicillium adametzii strain BIM-90. Mikologiya i Fitopatologiya, 34(4), 48–53.

41. Morozova, O. V., Shumakovich, G. P., Shleev, S. V., & Yaropolov, Y. I., (2007). Laccase-mediator systems and their applications: A review. Applied Biochemistry and Microbiology, 43(5), 523–535.

42. Muller, D., (1928). Studies on the new enzyme glucose oxidase. I. Biochem. Z, 199, 136–170.

43. Nakayama, M., Nakajima-Kambe, T., Katayama, H., Higuchi, K., Kawasaki, Y., & Fuji, R., (2008). High catalase production by Rhizobium radiobacter strain 2–1. Journal of Bioscience and Bioengineering, 106(6), 554–558.

44. Nema, A., Patnala, S. H., Mandari, V., Kota, S., & Devarai, S. K., (2019). Production and optimization of lipase using Aspergillus niger MTCC 872 by solid-state fermentation. Bulletin of the National Research Centre, 43(1), 1–8.

45. Nian, L., Cao, A., & Cai, L., (2020). Investigation of the antifreeze mechanism and effect on quality characteristics of largemouth bass

(Micropterus salmoides) during FT cycles by hAFP. Food Chemistry, 325, 126918.

46. Oksanen, J., Ahvenainen, J., & Home, S., (1985). Microbial cellulase for improving filterability of wort and beer. Journal of the Institute of Brewing, 91(3), 130–130.
47. Ough, C. S., (1975). Further investigations with glucose oxidase-catalase enzyme systems for use with wine. American Journal of Enology and Viticulture, 26(1), 30–36.
48. Panda, T., & Gowrishankar, B. S., (2005). Production and applications of esterases. Applied Microbiology and Biotechnology, 67(2), 160–169.
49. Pandey, A., Nigam, P., Soccol, C. R., Soccol, V. T., Singh, D., & Mohan, R., (2000). Review-advances in microbial amylases. Biotechnol. Appl. Biochem., 31, 135–152.
50. Pasha, K. M., Anuradha, P., & Subbarao, D., (2013). Applications of pectinases in the industrial sector. International Journal of Pure and Applied Sciences and Technology, 16(1), 89.
51. Patel, D. D., Patel, R. R., & Thakkar, V. R., (2015). Purification, characterization, and application of lipoxygenase isoenzymes from Lasiodiplodia theobromae. Applied Biochemistry and Biotechnology, 175(1), 513–525.
52. Payne, C. M., Knott, B. C., Mayes, H. B., Hansson, H., Himmel, M. E., Sandgren, M., & Beckham, G. T., (2015). Fungal cellulases. Chemical Reviews, 115(3), 1308–1448.
53. Polizeli, M. L. T. M., Rizzatti, A. C. S., Monti, R., Terenzi, H. F., Jorge, J. A., & Amorim, D. S., (2005). Xylanases from fungi: properties and industrial applications. Applied Microbiology and Biotechnology, 67(5), 577–591.
54. Puri, M., (2012). Updates on naringinase: Structural and biotechnological aspects. Applied Microbiology and Biotechnology, 93(1), 49–60.
55. Ramachandra, M., Crawford, D. L., & Hertel, G., (1988). Characterization of an extracellular lignin peroxidase of the lignocellulolytic actinomycete Streptomyces viridosporus. Applied and Environmental Microbiology, 54(12), 3057–3063.
56. Raveendran, S., Parameswaran, B., Beevi, U. S., Abraham, A., Kuruvilla, M. A., Madhavan, A., & Pandey, A., (2018). Applications

of microbial enzymes in the food industry. Food Technology and Biotechnology, 56(1), 16–30.

57. Regalado, C., García-Almendárez, B. E., & Duarte-Vázquez, M. A., (2004). Biotechnological applications of peroxidases. Phytochemistry Reviews, 3(1), 243–256.
58. Röcker, J., Schmitt, M., Pasch, L., Ebert, K., & Grossmann, M., (2016). The use of glucose oxidase and catalase for the enzymatic reduction of the potential ethanol content in wine. Food Chemistry, 210, 660–670.
59. Saadoun, I., Dawagreh, A., Jaradat, Z., & Ababneh, Q., (2013). Influence of culture conditions on pectinase production by Streptomyces sp. (strain J9). International Journal of Life Science and Medical Research, 3(4), 148.
60. Sadot, M., Curet, S., Chevallier, S., Le-Bail, A., Rouaud, O., & Havet, M., (2020). Microwave-assisted freezing; Part 2: Impact of microwave energy and duty cycle on ice crystal size distribution. Innovative Food Science and Emerging Technologies, 62, 102359.
61. Sanghi, A., Garg, N., Gupta, V. K., Mittal, A., & Kuhad, R. C., (2010). One-step purification and characterization of cellulase-free xylanase produced by alkalophilic Bacillus subtilis ash. Brazilian Journal of Microbiology, 41(2), 467–476.
62. Schülein, M., (1988). Cellulases of Trichoderma reesei. Methods in Enzymology, 160, 234–242.
63. Shah, N. P., Spurgeon, K. R., & Gilmore, T. M., (1993). Use of dry whey and lactose hydrolysis in yogurt bases. Milchwissenschaft, 48(9), 494–498.
64. Sharma, R., Chisti, Y., & Dan Banerjee, (2001). Production, Purification, Characterization and Applications of Lipases. Biotechnology Advances, 19, 627–662.
65. Sindhu, R., Binod, P., Mathew, A. K., Abraham, A., Gnansounou, E., Ummalyma, S. B., & Pandey, A., (2017). Development of a novel ultrasound-assisted alkali pretreatment strategy for the production of bioethanol and xylanases from chili post-harvest residue. Bioresource technology, 242, 146–151.
66. Singh, R. S., Sooch, B. S., & Puri, M., (2007). Optimization of medium and process parameters for the production of inulinase from a newly isolated Kluyveromyces marxianus YS-1. Bioresource Technology, 98(13), 2518–2525.

67. Singh, R., Jha, A. B., Misra, A. N., & Sharma, P., (2020). Entrapment of enzyme in the presence of proline: effective approach to enhance activity and stability of horseradish peroxidase. 3 Biotech., 10(4), 1–8.
68. Sinnott, M. L., (1990). Catalytic mechanism of enzymic glycosyl transfer. Chemical Reviews, 90(7), 1171–1202.
69. Soccol, C. R., Rojan, P. J., Patel, A. K., Woiciechowski, A. L., Vandenberghe, L. P., & Pandey, A., (2006). Glucoamylase. In Enzyme Technology, 2(6), 221–237.
70. Sukumaran, R. K., (2009). Bioethanol from lignocellulosic biomass. Part II: Production of cellulases and hemicellulases. Handbook of Plant-Based Biofuels (pp. 141–157). Boca Raton, FL, USA: CRC Press, Taylor & Francis Group.
71. Tanrıöven, D., & Ekşi, A., (2005). Phenolic compounds in pear juice from different cultivars. Food Chemistry, 93(1), 89–93.
72. Tapre, A. R., & Jain, R. K., (2014). Pectinases: Enzymes for fruit processing industry. International Food Research Journal, 21(2).
73. Wang, W., Sun, M., Liu, W., & Zhang, B., (2008). Purification and characterization of a psychrophilic catalase from Antarctic Bacillus. Canadian Journal of Microbiology, 54(10), 823–828.
74. Wei, W., Wang, R., Nie, L., & Yao, S., (1997). Rapid determination of dimethoate with a surface acoustic wave impedance sensor system. Analytical Letters, 30(15), 2641–2653.
75. Xiang, H., Yang, X., Ke, L., & Hu, Y., (2020). The properties, biotechnologies, and applications of antifreeze proteins. International Journal of Biological Macromolecules, 153, 661–675.
76. Zhang, B., Cao, H. J., Wei, W. Y., & Ying, X. G., (2020). Influence of temperature fluctuations on growth and recrystallization of ice crystals in frozen peeled shrimp (Litopenaeus vannamei) pre-soaked with carrageenan oligosaccharide and xylooligosaccharide. Food Chemistry, 306, 125641.
77. Zhao, C., Su, H., & Liu, Y., (2016). Catalytic mechanism of acetolactate decarboxylase from Brevibacillus brevis towards both enantiomers of α-acetolactate. RSC Advances, 6(84), 80621–80629.
78. Zhao, F., Wang, Q., Dong, J., Xian, M., Yu, J., Yin, H., & Wang, J., (2017). Enzyme-inorganic nanoflowers/alginate microbeads: An enzyme immobilization system and its potential application. Process Biochemistry, 57, 87–94.

79. Zhao, X., Shi-Jian, D., Tao, G., Xu, R., Wang, M., Reuhs, B., & Yang, Y., (2010). Influence of phospholipase A2 (PLA2)-treated dried egg yolk on wheat dough rheological properties. LWT-Food Science and Technology, 43(1), 45–51.
80. Zhu, Y., Jia, H., Xi, M., Li, J., Yang, L., & Li, X., (2017). Characterization of a naringinase from Aspergillus oryzae 11250 and its application in the debitterization of orange juice. Process Biochemistry, 62, 114–121.

Chapter 7

Role of Microbial Fermentation in Food Quality Enhancement

CONTENTS

7.1. INTRODUCTION

As we know that fermented foods are healthier than unfermented foods (Hasan et al., 2014). The fermenting microorganisms present in fermented foods enhance the nutritional value and the fermentation by microorganisms occur by three different methods, and these are discussed below.

Microorganisms are both anabolic and catabolic, produce complex vitamins and other growth elements and decompose complex compounds. (Kennedy, 2016). The nutrients that are stuck in the cells and plant structures are released by indigestible elements. The grains and single seeds usually have this event. During the process of milling, to liberate nutrients, the physical division of hemicellulosic and cellulosic structures covering the endosperm (i.e., digestible carbohydrates and having a lot of proteins) has occurred. To take out nutritional elements, crude milling is used in underdeveloped areas, but from the plant products, it is inadequate to liberate full nutritional value. Some of the bounded nutrients do not reach to the human digestive system, even with the process of cooking. Simultaneously, by specific molds, bacteria, and yeasts, this problem can be fixed as they breakdown the indigestible coverings and cell walls of these foods both chemically and physically (Figure 7.1) (Hasan et al., 2014).

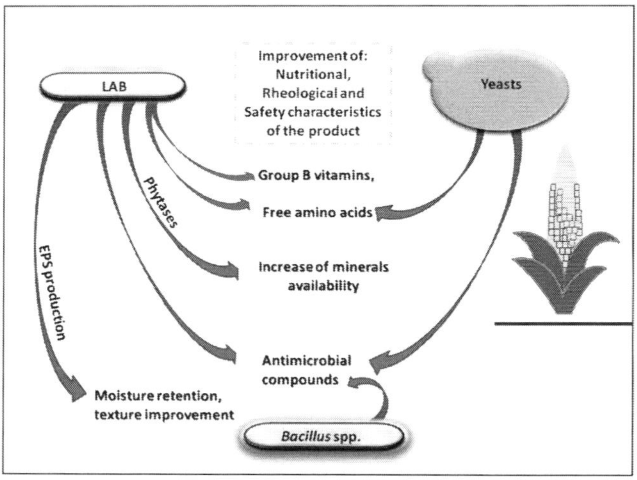

Figure 7.1. On the characteristics of the ending fermented maize product, the influence of microbial activities.

Source: https://www.mdpi.com/2311-5637/6/1/36/htm.

The enzymatic decomposition of polymers that are indigestible by humans into simple sugars and their derivatives like hemicelluloses, cellulose, and the same kind of polymers is one of the methods to enhance nutritional properties of the plant material. For human intake, the enhancement in the cellulose-containing substances of fermented foods can be done by using microbial enzymes (Sharma et al., 2010). Several cereal foods have less nutritional components and are ingested as an important diet for poor people. Though, in order to increase food digestibility and nutritional content, yeast fermentation and L.A.B. were used. The microbial enzyme action is increased by fermentation method as it delivers an acidic atmosphere at a temperature range of 22–25°C. The bioavailability of proteins (proteases), simple sugars or polysaccharides (amylases), iron, and free fatty acids (lipases) is improved by a decrease in anti-nutrients levels viz. phytic acid (decomposition by using phytases) and tannins and it is included in the crucial function of enzymatic hydrolysis in fermented foods (Table 7.1 and Figure 7.2).

Table 7.1. Few Frequently Prepared Fermented Beverages/Foods and Their Fermenting Microorganisms

Fermented Beverages/Foods	**Substrates Used**	**Fermenting Microorganisms**
Cereals: Chilra, Bhatura, Marchu, Ambali, Dosa, Kunu-Zaki	Rice, wheat, maize, sorghum, millet	L. pantheris, L. plantarum, Penicillium sp., S. cerevisiae, L. mesenteroides, E. faecalis, Trichosporon pullulans, Pediococcus acidilactici, P. cerevisiae, Delbrueckii hansenii, Deb. tamari
Beverages: Sake, Wine, Kombucha, Beer	Cereals, grapes, rice	L. sakei, L. curvatus, L. plantarum, Leuconostoc carnosum, Leuconostoc gelidium, B. licheniformis, E. faecalis, E. hirae, E. durans, Bacillus subtilis, L. divergens, L. carnis, E. cecorum, B. lentus
Vegetable Products: Kimchi, Tempeh, Natto, Miso, Sauerkraut	Ginger, soybean, cucumber, cabbage, radish, broccoli	Leuconostoc mesenteroides, Aspergillus sp., Rhizopus oligosporus, R. oryzae, L. sakei, L. plantarum, Thermotoga sp., L. hokkaidonensis, L. rhamnosus, Rhodotorula rubra, Leuconostoc carnosum, Bifidobacterium dentium, E. faecalis, Weissella confusa, Candida sake
Meat Products: Salami, Sucuk, Nham, Arjia, Jama	Meat	Lactobacillus bulgaricus, Lactococcus lactis, L. acidophilus, L. cremoris, L. casei, L. paracasei, L. thermophilus, L. kefiri, L. caucasicus, Penicillium camemberti, P. roqueforti, Acetobacter lovaniensis, Kluyveromyces lactis, Saccharomyces cerevisiae
Dairy Products: Yogurt, Kefir, Curd, Cheese, Yakult	Milk casein and milk	Aspergillus oryzae, Zygosaccharomyces bailii, S. cerevisiae, A. pasteurianus, Gluconacetobacter, A. xylinus, Komagataeibacter xylinus

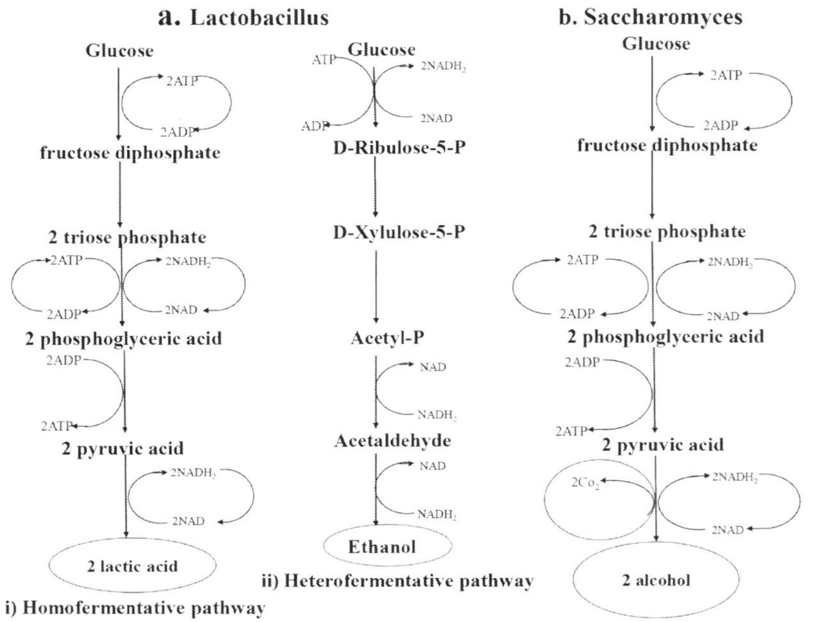

Figure 7.2. To show yeast and L.A.B., the sugar breakdown by Saccharomyces and Lactobacillus.

Source: https://www.semanticscholar.org/paper/Microbial-Fermentation-and-Its-Role-in-Quality-of-Sharma-Garg/21f0d7a72d941b78e5575c0027382e784a676db9.

7.2. EFFECTS OF LACTIC ACID (LA) FERMENTATION ON THE NUTRITIONAL ASPECTS OF FOOD

The number of important nutrients present and digestibility are two main factors that take part in food's nutritional value. The process of fermentation may enhance digestibility as well as nutrients. At the start the macronutrients may be digested by the enzymes of fermented microorganisms in fermentation process (Xiang et al., 2019). Fermentation can affect the nutritional quality of food in numerous ways, which includes improving nutrient density and increasing the bioavailability and quantity of nutrients. The former may be obtained by making promoters for the decomposition and absorption of anti-nutritional factors, pre-digestion of different food elements, affecting the uptake of nutrients by the mucosa (Nkhata et al., 2018). Lactic acid (LA) fermentation process improves the availability of some micronutrients, limiting amino acids and the solubility of proteins (Rollán et al., 2019). Phytates, tannins (50%), and oligosaccharides (90%) are also decreased by this process. Nutritional diseases can have an indirect or direct nutritional effect of fermented foods. There is a direct curative impact on fermentation process of food. Similarly, by enhancing the availability of vitamins such as folic acid, niacin, riboflavin, thiamine, food fermentation has a direct impact on consumers' health. By the decomposition of complex elements into inorganic iron with vitamin C, it also improves iron utilization (Figure 7.3) (Nkhata et al., 2018).

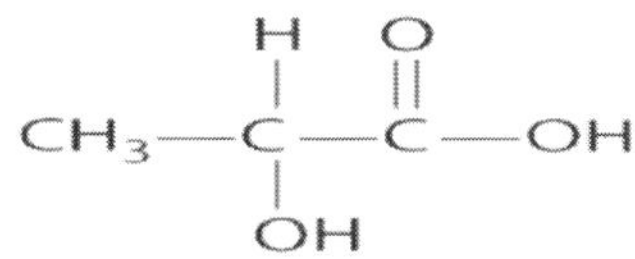

Lactic acid

Figure 7.3. Lactic acid formula.

Source: http://www.msrblog.com/science/chemistry/about-lactic-acid.html.

By decreasing non-digestible substances in plants such as polygalacturonic acids and glucuronic, hemicelluloses, and cellulose, food fermentation enhances trace element and bioavailability of minerals. By limiting cholesterol absorption in the intestine and by endogenous cholesterol synthesis in the dietary and liver, it also decreases serum cholesterol. It is stable, strong, and safe for the substance, hence preventing infections/diseases such as salmonellosis and diarrhea (Minh, 2014).

7.3. ENRICHMENT AND CHANGES OF BIOLOGICAL COMPONENTS IN FERMENTED FOODS

7.3.1. Vitamins Bio-Enrichment

In a few synthesized, selected foods, nutrients, mostly vitamins, are labeled as a public health measure; for example, ascorbic acid (vitamin C) can be added in fruit juices, while in making bread, vitamin D is added to riboflavin and milk (Figure 7.4). Though, because of its high-cost value, only Western world can use this fortification process. Therefore, for the biological enhancement of foods, numerous countries should use this kind of food fermentation. The refined white rice causes deficiency of thiamine (Vitamin B1). A disease called beriberi cause paralysis and strokes, can be caused by this kind of rice. There can be a sudden death of babies at three months due to heart failure if they are fed by thiamine-deficient (leads to beriberi) mothers (Nout and Kiers, 2005). The microorganisms included in the tape Ketan fermentation produce thiamine. The recovery of the thiamine level found in unpolished rice is also done by these microorganisms. Hence, this can help the rice-eating person.

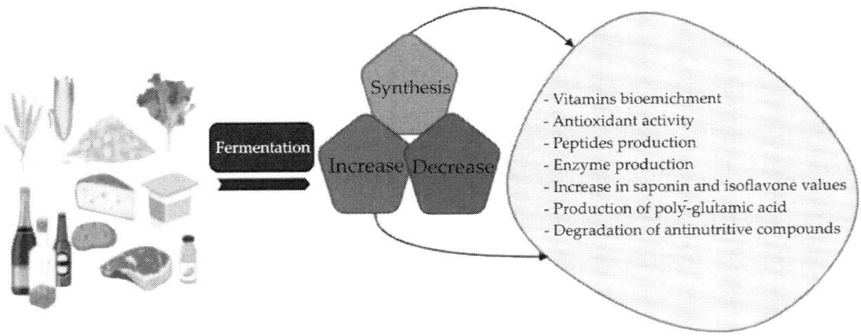

Figure 7.4. Nutritional improvement in fermented foods.

Source: https://www.mdpi.com/2311-5637/6/4/106/htm.

By marinating, removing hull from seeds, and half-cooked soya beans using Rhizopus oligosporus or other alike patterns, "Tempe" that is an Indonesian origin-rich dish, is prepared (Nout and Kiers, 2005). By joining cotyledons into slices and being cooked, a solid cake is prepared by this mold. In the fermentation process, the proteins are half hydrolyzed; the lipids get breakdown to their components like stachyose (humans can digest a tetrasaccharide), niacin increases seven times, fatty acids, riboflavin increases by two folds, and vitamin B-12 is not present in vegetarian food. Vitamin B-12 is produced by an essential mold growth of fermenting bacterium and it is commonly not present in vegetarian foods (Nout and Kiers, 2005). The texture of many bowls of cereal/legume mixtures and the digestibility enhance and the cooking time decreases by the formation process of tempe. When injected into idli fermentation of India, vitamin B-12 is formed by the bacterium, i.e., Klebsiella pneumoniae (nonpathogenic strain).

The oldest alcohol-holding beverage on the American continent and Mexican plaque is produced by the fermentation of the cactus plant

(Agave) juice. Being rich in thiamine, niacin, riboflavin, pantothenic acid, and pyridoxine and p-aminobenzoic acid biotin, poor people of Mexico typically intake plaque (Sharma et al., 2020). There is a good sour taste and a thin gruel constancy of an alcoholic beverage like Kaffir beer. The people of Bantu of South Africa traditionally produce Kaffir beverage with 1 to 8% concentration of alcohol. Unmalted and malted kaffir corn (Sorghum caffrorum) were used to prepare Kaffir beverage. Maize or millet can be the replacement of kaffircorn. In people eating maize, at the time of fermentation, this alcoholic beverage enhances the level of niacin/nicotinic acid and riboflavin nearly two times by keeping the thiamine level constant (Aka et al., 2014).

Palm sap is a translucent liquid with no color, and it has almost 10 to 12% fermentable sugar, it is a plump, sweet, and translucent white suspension of yeasts and bacteria. This kind of wine contains almost 83 mg of ascorbic acid/L. There is an increment in pyridoxine from 4–18 μg/L, thiamine from 25–150 μg/L, and riboflavin from 35–50 μg/L in fermented palm wine. Astonishingly, palm wine has a large amount of vitamin B-12 (190 to 280 μg/mL). The inexpensive sources of vitamin B are palm toddies and play a critical role in nutrition thriftily drained in the tropics (Anal, 2019).

7.3.2. Antioxidant Activity

The decrease in 2^0-azino-bis (3-ethylbenzo-thiazoline-6-sulfonic acid; A.B.T.S.), power assay, total phenol content (TPC) prediction and 1,1-diphenyl-2-picryl hydrazyl (D.P.P.H.) radical scavenging activity are the antioxidant activities in the fermented foods (Abubakr et al., 2012). There are some antioxidant properties in many Asian soybean fermented foods for example, bekang (fermented soybean foods of India), and tungrymbai, kinema (fermented soybean food of Nepal and India), thuanao (fermented soybean food of Thailand), natto (fermented soybean food of Japan), tempe (mold-fermented soybean food of Indonesia), jang, and chungkokjang (fermented soybean foods of Korea, and douchi (fermented soybean food of China). Kimchi and yogurt also has antioxidant activity (Astuti et al., 2013). At the time of fermentation of skim milk, the antioxidant activity of LA bacteria was found by Abubakr et al. (2012), with the help of 1,1-diphenyl-2-picrylhydrazyl (D.P.P.H.) assay and noticed free radical searching activity varying between 14.7 to 50.8% (v/v) after 24 to 72 hours fermentation, respectively (Abubakr et al., 2012). It has been observed that the antioxidant activity was greatly ($p \leq 0.05$) affected by fermentation time and the different classes of L.A.B. used evaluated the enhancement in the quality of fermented sausage concerning antioxidant value by L. curvatus (SR6) and L. paracasei (SR10-1) (Zhang et al., 2017). The bacteria greatly improved the sausage's antioxidant activity, which results in the overall enhancement in the sausage safety and sensory attributes. In the case of L. curvatus SR6 viz. 47.31% ± 4.62% and 59.67% ± 6.68% decreasing power and the D.P.P.H. scavenging activity was better whereas with L. paracasei SR10-1, i.e., 63.89% ± 0.93% and 285.67% ± 2.00%, anti-lipid peroxidation capacity and O.H. scavenging activity were better, respectively (Dajanta et al., 2013).

7.3.3. Peptides Production

At the time of food fermentation, proteolytic microorganisms (mostly Bacillus) produce the bioactive peptides (B.A.P.s). Peptides also play an important role as immunomodulatory agents and antithrombic and contain some antihypertensive properties (Hou et al., 2015; Alcántara et al., 2016). The production of B.A.P.s is less than the benchmark due to rising costs of the enzymes for protein hydrolysis and unavailability of large-scale manufacturing facilities. Though, for the creation of B.A.P.s, microbial fermentation is a more cost-effective and reasonable mechanism. The B.A.P.s have health properties related with dairy products. Until now, cell envelope proteinase (C.E.P.s) of four different kinds have been found, which are PrtP from L. casei and L. paracasei, PrtB from L. delbrueckii subsp. bulgaricus, PrtH from L. helveticus and PrtR from L. rhamnosus, and L. plantarum. whereas the presence of four different paralogs of PrtH (i.e., PrtH1, PrtH2, PrtH3, and PrtH4) has been discovered in L. helveticus, and their distribution is strain-dependent but usually, most of the Lactobacillus species have only one C.E.P. (Miyamoto et al., 2015). Amongst the Lactobacillus genus, which is only responsible for the formation of many varieties of B.A.P.s, L. helveticus is the most proteolytic class due to numerous C.E.P. paralogs and different specifications (Qian et al., 2011).

In some milk products fermented earlier like yogurt, kefir, koumiss, fermented products of fish cheese, and fermented milk of camel, the inhibition activity by angiotensin-converting enzyme (A.C.E.) has also been noticed. Fermentation of soybean products produce bioactive peptides, which helps treat and stop metabolic diseases (Moslehishad et al., 2013; Sanjukta and Rai, 2016).

7.3.4. Enzymes Production Through Microorganisms

Bacillus, in Asian soybean, a fermented food product that breakdown complex substances into simple biomolecules, is fermenting microorganisms that typically produce enzymes like proteinase, amylase, catalase, mannase, cellulose, etc. (Table 7.2). Mycelia fungi like Amylomyces produce carbohydrates forming enzymes like β-galactosidase, amyloglucosidase, pectinase, α-amylase, invertase, maltase, cellulase, lipase, and alkaline proteases. Fermented beverages/foods have Aspergillus, Monascus, Actinomucor, Rhizopus, Mucor, and Neurospora. There are several uses of enzyme produced by A. oryzae in koji, i.e., Taka α-amylase (T.A.A.) in industries. In order to synthesize alcohol, a dry, steady, and cake-like amylolytic starter cultures is used in the Himalayan region. To increase amylase, yeast strains are mixed with these pioneer cultures such as Pichia burtonii, Saccharomycopsis capsularis, and S. fibuligera, (Tsuyoshi et al., 2005; Tamang et al., 2007).

Table 7.2. Few Important Commercial Enzymes Used in Fermented Beverages/ Foods

Substrates	Enzymes	Microbial Source	Enzymatic Action/ Process
Cereals	Pentosanase Lipases Amylase Protease Phytase Pullulanase Glucose oxidase Xylanase β-glucanase	P. notatum, A. niger B. licheniformis and B. subtilis B. subtilis, A. niger and P. funiculosum A. niger Trichoderma sp. A. oryzae and B. subtilis Aspergillus niger B. acidopullulyticus	Liquefaction, mashing, malting, and flavor, esters produced
	Amyloglucosidase	A. niger and A. flavus	–
	Pectinase	A. niger	–
	A-acetolactate-decarboxylase	B. subtilis	–
	Cellulase	T. longibrachiatum	–
Meat	Papain protease	A. niger, S. aureus A. oryzae, T. longibrachiatum, and B. subtilis	Soften the meat
Dairy	Catalase Lactase Protease	B. subtilis A. niger, A. orzyae and B. subtilis S. boydii and Bacillus sp.	Lactose-free milk El cheese formation Iminating H_2O_2
Beverages	Tannase Glucose oxidase	A. niger P. notatum	Eliminating O_2 hydrolysis of esters Explanation of juices

For fibrinolytic activity of the enzyme nattokinase synthesized by B. subtilis present in natto, it has been noticed. Fibrinolytic activity is also exhibited by other bacterial strains isolated from fermented foods like Vagococcus carniphils, B. amyloliquefaciens, P. acidilactici, V. lutrae, Enterococcus faecalis, E. gallinarum and E. faecium. Fibrinolytic activity was also showed by the SK1-3-7 strain of Virgibacillus halodenitrificans isolated from fermented fish sauce (Zeng et al., 2013; Montriwong et al., 2015).

7.3.5. Increase in Saponin and Isoflavone Values and Poly-Glutamic Acid Production

Soybeans: aglycones, β-glucoside, acetylglucoside, and malonylglucoside have four kinds of chemicals which derive isoflavones such as glycitein, genistein, and daidzein. Few Asian fermented products of soybean, like sufu and douche-China, miso, and natto-Japan, chungkokjang-Korea, thuanao, and doenjang-Thailand and tempe-Indonesia break isoflavone glucosides into their respective aglycones in fermentation. At the time of fermentation of tempe, the ingredients of isoflavone and aglycone, particularly Factor-II, were discovered to enhance (Nakajima et al., 2005). The LDL-C receptor activation, have ability to avoid vascular diseases, enhance by observing isoflavones in doenjang (Kwak et al., 2012).

There are additional two kinds of Soybean saponins which are naturally oleanane triterpenoid glycosides, i.e., Group A and D.D.M.P. (2,3-dihydro-2,5-dihydroxy-6-metyl-4H-pyran-4-one). There are some useful health effects of D.D.M.P. and their derivatives, Groups B and E saponins like inhibition of hypercholesterolemia, elimination of the proliferation of colon cancer cells, prevention from peroxidation activity of lipids. Natto's saponin content has some increment. The high content of Kinema in Group B saponins has been discovered, representing its goodness to consumers (Paucar et al., 2010).

Strains of Bacillus sp., like B. subtilis and B. licheniformis in Asian fermented soybean products form Poly glutamic acid (P.G.A.) and is not produced by proteins present in ribosomes (Murata, 2005). Viscous materials of Asian fermented soybean products are safe to ingest due to their water-soluble properties, biodegradable, and non-poisonous nature towards humans.

7.3.6. Anti-Nutritive Compounds Degradation

In most of the food substrates, anti-nutritive substances are discovered. By reducing the nutritive value of food, they restrict the availability of nutrients to the body, and they are poisonous to human beings. Microorganisms occupied in fermented foods may decrease anti-nutritive substances, making inconsumable products safe for ingesting (Tamang et al., 2016). Other steps employed during processing are, such as help minimize the cyanide content in the finished products in several African fermented cassava products, gari, and fufu, in order to reduce the amount of cyanide, few other steps are taken at the time of process like dewatering, grating, fermentation, washing, roasting, and peeling. In some variations of cassava tubers that are bitter, the existence of cyanogenic glycoside lotaustralin and linamarinis is detoxified during traditional ways of fufu and gari production by Lactobacillus, Streptococcus, and Leuconostoc to synthesize hydrocyanic acid (H.C.N.) with a low boiling point that releases during toasting from the dewatered pulp producing the final product nontoxic and safe for usage (Bamidele et al., 2015). Though, to eliminate the flatulence from indigestible oligosaccharides such as stachyose and verbascose and transform it into absorbable disaccharides and monosaccharides, Tempe uses R. oligosporus. While B. subtilis reduces anti-nutritive compounds in kinema, rabadi, and idli fermentation, the phytic acid is also decreased.

7.3.7. Biochemical Changes During Cereal Fermentation

All over the world, cereals are the most vital vitamins, carbohydrates, minerals, dietary proteins, and fibers. Though, in terms of sensory properties and nourishing quality of the products related to milk and milk products, the acceptance of cereals amongst consumers is the main issue with cereals. This is because of the low availability of starch, deficiency of some vital amino acids (for example, lysine), low amount of proteins, the rough nature of the cereals and the presence of anti-nutrients (such as polyphenols tannins, and phytic acid) (Verni et al., 2019).

To enhance the nutritional amount of cereals, various approaches have been applied. For example, improvement through the change of rich sources of proteins such as defatted/legumes oilseed meals of cereal grains or supplementation of amino acids with essences of proteins and genetic modifications. In addition to this, the nutritive quality of cereals, among which fermentation is the best one, is increased by several processing technologies such as fermentation, sprouting, milling, and cooking. Usually, the production of amino acids and vitamin B group availability improves, and the level of carbohydrates with polysaccharides and non-digestible oligo reduces by natural fermentation of cereals. By providing optimal pH conditions present in a complex form with polyvalent cations such as zinc, magnesium, iron, and calcium, natural fermentation enhance enzymatic breakdown of phytates. The availability of soluble iron, calcium, and zinc by large folds can be increased by a decrease in phytate (Mariod et al., 2016).

7.3.8. Presence of Biogenic Amines in Juices and Vegetables Fermented with Lactic Acid (LA) Bacteria

Biogenic amines (B.A.) consist of amines with low molecular mass and good biological efficacy and are mostly present in beverages and food items (Biji et al., 2016). Numerous psychoactive effects and/or vasoactive symptoms are due to over-intake of B.A.

At the time of impulsive or controlled fermentation, food decomposition usually result in an enhancement in the concentration of B.A., where factors like sodium chloride content, temperature, oxygen, pH, badly affect the production of B.A (Sharma et al., 2020).

In several European countries, for the sensorial properties and nutritional value, shredded white cabbage, commonly called sauerkraut, is preserved by LA fermentation. The microorganisms that synthesize B.A., show three steps through which Sauerkraut is produced. Some of these are like Lactobacillus sp., Leu. mesenteroides, and P. cerevisiae.

The mean concentrations of B.A. present were detected to be 146, 50, and 174 mg/kg for putrescine, cadaverine, and tyramine, respectively in 121 samples of sauerkraut in a study showed by Kalac et al. (2000). In 19% samples, the Histamine values was above 10 mg/kg, whereas it was less than 2 mg/kg in 44% samples. The study by Karovicova and Kohajdova showed that pasteurized and bottled sauerkraut juices (Germany) have a large quantity of putrescine up to 694 mg/dm^3. Another study carried out by Kalac et al. (2000), in which sauerkrauts was prepared after 6 months of storage in different laboratory and the quantity of cadaverine, tyramine, spermidine, and putrescine were determined. In LA fermented cabbage juices of Germany, the same study was carried out for B.A.'s determination (Kirschbaum et al., 2000). The concentrations of histamine, cadaverine, putrescine, spermidine, and tyramine were determined in the range of 1–15 mg/kg in the carrot and red beet, synthesized by LA fermentation. (Hornero et al., 1997).

With specific conditions such as pasteurization, acidity, and pH, reached 9–10 g/kg and 3.8–4.0, respectively, and temperature of 15–20°C for initial fermentation, Kalac et al. prepared sauerkraut with less amount of B.A. to reserve them from bacteria. Kolesarova (1995) also discovered that the quantity increases with pH values ranging from 3.6–3.8, therefore recommending when the pH goes below 4, pasteurization is stopped, as at this point yeast activity such as P. cerevisiae is initiated, raises the quantity of histamine to 200 mg/kg. Another study carried out by Kalac et al. observed that sauerkrauts inoculated with mixed culture of L. casei, E. faecium Pediococcus sp., L. plantarum lead or single L. plantarum results in the formation of low concentrations of tyramine, putrescine, and cadaverine.

The manufacturing of machines for silos, shredding, and transporters and the early stoppage of contamination by monogenic bacteria from shredded cabbage, are done by the factors that involve in the reduction of B.A levels in sauerkraut production. Though, the resistance of amine-negative inoculants to contamination was also described by Kalac et al. (2000).

7.4. NUTRITIONAL VALUE OF FERMENTED DAIRY PRODUCTS

Due to more nutritional content related to simple milk, the market values of fermented dairy products viz. yogurt, kumiss, sour milk, dahi, acidophilus milk, and other similar milk products is much higher. The configuration of minerals remains the same while the quality and quantity of vitamins, fat, proteins, and carbohydrates vary in fermented milk (Figure 7.5). The elements formed at milk souring's biochemical reactions and the fermenting microorganisms control the quality of sour milk. The substances are like LA, alcohol, carbon dioxide (CO_2), antibiotics, and vitamins.

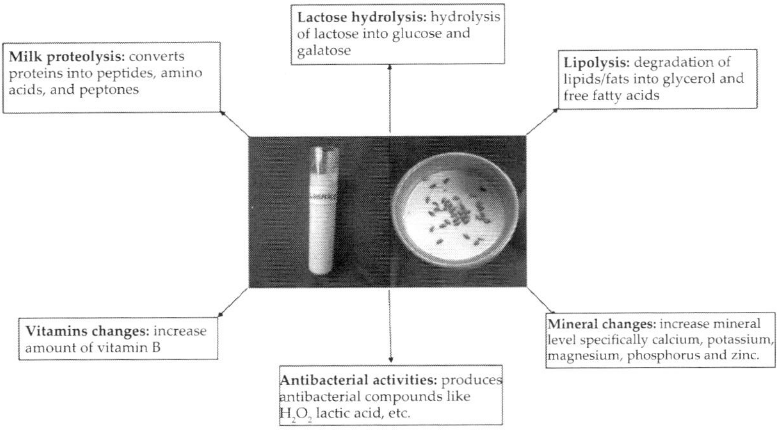

Figure 7.5. During fermentation, variations in fermented dairy products.

Source: https://www.mdpi.com/2311-5637/6/4/106/htm.

For enhancing the nutritional value of fermented milk, the useful biochemical processes are presented in subsections.

7.4.1. Milk Proteolysis

The breakdown of proteins into peptones peptides, and amino acids is led by proteolysis in milk, which suggest an increased quantity of amino acids like leucine, methionine, phenylalanine isoleucine, threonine, tryptophan, valine, and tyrosine. These are vital A.A.s giving certain benefits, particularly to physically weak people. Endo peptides and Exo of L.A.B. help in the process of proteolysis (Hao et al., 2020). In fermented milk such as yogurt, dahi, kefir, etc., the quantity of protein rises from 85.4 to 90% and due to the precipitation of milk by LA into fine curd which has higher digestibility for proteins, causing increase in nutritional value. At the time of storage and fermentation of fermented milk products, the quantity of free amino acids, particularly phenylalanine, isoleucine, proline, lysine, arginine, and cysteine, milk products become very nutritional due to biochemical fluctuations in milk protein and proteolysis in milk. (Savijoki et al., 2006).

7.4.2. Lactose Hydrolysis

Few bacteria present in the milk cause the lactose hydrolysis in milk. Related to an average 5% lactose in milk, the hydrolysis of lactose synthesizes 16–20% galactose, 45–50% lactose and 0.6–0.8% glucose. The hydrolysis of lactose is done by L.A.B. through the formation of β-galactosidase. For LA formation, hydrolysis of lactose is essential, which prevents the putrefaction of microorganisms from growing and lessens the bowel's pH. Moreover, for the absorption of calcium and organoleptic properties, LA is important (Dutra et al., 2017).

7.4.3. Lipolysis

The size of the fat globules reduce and become digestible by the process of homogenization. Because of the increment in free fatty acid content by LA bacteria, lipolysis leads to physiological effects.

7.4.4. Vitamins Changes

In fermented milk, the components of vitamin depends on the bacterial culture present in it. With the growth of bacteria in milk, ascorbic acid, vitamins B_1, B_2 become half in quantity while most vitamin B groups, especially thiamine, riboflavin, and nicotinamide, are increased two times through consumption (Yoshii et al., 2019).

7.4.5. Antibacterial Activity

On the basis of the compounds making antibacterial properties like LA, hydrogen peroxide, antibiotics, and bacteriocins and the growth of bacteria's antibiotic activity, like lactobacilli in yogurt, the bactericidal action in fermented milk takes place (Vieco et al., 2019).

7.4.6. Mineral Changes

During and after fermentation, LA bacteria causes the increase and higher bioavailability of minerals in fermented milk, particularly in potassium, calcium, magnesium, zinc, phosphorus, and potassium iodide, as a result of a process of acidity and fermentation (Bintsis, 2018).

7.5. BIOCHEMICAL CHANGES IN MEAT-BASED FERMENTED FOOD PRODUCTS

Few facultative/obligate anaerobes such as Streptococcus, Pediococcus, Lactobacillus, Enterococcus, Leuconostoc, and Lactococcus are related to acidogenic (LA) and gram-positive bacteria and with different points of efficiency. These bacteria are used to break down saccharides into alcohols, lipids, amino acids, LA, and aliphatic compounds (Figure 7.6) (Bintsis, 2018).

In fermented sausages, three purposes are performed by these organisms at the same time, i.e., when combined with myoglobin, responsible for the cured color, by generating DL-LA from glucose through anaerobic glycolysis, decreases pH and making nitric oxide by decreasing nitrate and nitrite. (Mastanjević et al., 2017).

Decrease in pH, transformation of simple sugars into LA and fast growth of the L.A.B. are due to the sausage incubation at optimal temperature with facultative anaerobic conditions. To decrease the pH, a postmortem range of 4.5–7 μmol/g is not adequate; so, as a substrate for L.A.B, put in simple sugars, taking pH to 4.6–5. To decrease the pH by 0.1, 0.62 g glucose/kg of meat was used by Reference (Mastanjević et al., 2017).

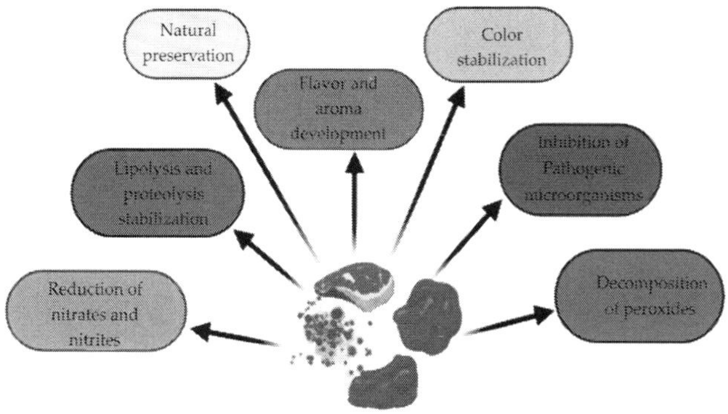

Figure 7.6. Quality fluctuations in fermented meat products.

Fermentation of carbohydrates combine with an adequate amount of phosphorylation is the initial method for the breakdown of LA.

The Embden-Meyerhof-Parnas Pathway with LA is the pathway of homofermentative LA fermentation that results in LA as the only end product, giving a sharp, tangy taste. Phosphoketolase/6-phosphogluconate pathway also produces LA by using heterofermentative LA bacteria, but a small quantity of acetic acid (AA) (approximately 10%) is also released, which is one of the reasons of an off taste from hexoses (Papagianni, 2012).

Relative humidity, time, and temperature are vital factors in LA fermentation; for instance, temperature, and higher water activity (a_w) help in decreasing the pH level and rapid growth of L.A.B. Smoke can add thousands of antimicrobial compounds and smell like phenols (antioxidants), organic acids (e.g., acetic, isobutyric, formic, propionic acids, and butyric), and carbonyls taking part in the hindrance of microorganisms and coagulation of surface proteins (Longhi et al., 2013).

The transformation of lactate and ammonia into LA determines the fluctuation in pH. The carbohydrates produced from bacterial glycerol fermentation and ammonia generated from fermentation of amino acid add to carry out this formation. In this process of fermentation, AA is also produced. Moreover, oxygen used via metabolic process and other settings affect the kind of microorganisms and their metabolism, which influence the formation of lactose and the quantity of fermented carbohydrates (Juturu and Wu, 2016).

Myofibrillar proteins break down into polypeptides with low pH with the aid of Cathepsin D (muscle's endogenous enzyme). Non-volatile and volatile flavors in fermented sausages are produced by protein proteolysis, leading to peptides and free amino acids. The breakdown of flavor- and aroma-generating compounds like sarcoplasmic and myofibrillar proteins is carried out with the aid of L. plantarum and L. casei, is based on the literature quoted by researchers. These myofibrillar proteins are more important than sarcoplasmic proteins for the creation of the structure of fermented sausages.

Microbial enzymes of muscle tissues cause lipolysis in fermented meat products, i.e., both exo- and endoenzymes, and the oxidation of fatty acids leads to the formation of aldehydes, alkanes, alcohols, and ketones (Pasini et al., 2018).

Oxidative and hydrolytic variations are autocatalytic and usually linked with unsaturated fats. They are the cause of the release of flavors and aromas (both bad and good flavors). In non-vacuum conditions, thiobarbituric acid reactive substances (T.B.A.R.S.) increases minutely, while in vacuum packages, it remains fixed. Lipolytic enzymes release fatty acids, and it is naturally present in meat while the activity is slight in semi-dry fully cooked products. By reducing light penetration and direct usage, oxygen can decrease by using mold ripening and smoking.

Biochemical reactions are affected by sausage diameter due to less oxidation; greater diameter sausages (anaerobic) does not influence the maximum lipase activity and is not affected by the concentration of glucose; though, it is vital for fast lipase formation (Karsloğlu et al., 2014).

Up till now, from fermented meat items, 200 volatile compounds have been found that generate aroma. Acetoin, diacetyl (0.1 ppm), smoking, spices, acetaldehyde (0.7 to 1.5 ppm), and carbohydrate metabolism also produce these compounds. For example, butter-like aroma within the range of 1,300–1,400 ng/g is produced by diacetyl and acetoin, whereas AA produces vinegar aroma. Phenylalanine degrades into aldehydes, and malty and fruity aroma is generated by methyl aldehyde, whereas methyl acids have a cheesy and sweaty aroma. Breakdown of lipids makes straight-chain aldehydes alkanes, and alkenes, fruity, metallic, and rancid in flavor and green in color. Methyl ketones are combined with Staphylococcus carnosus and produce fruity, musty, and cheese-like aroma, whereas 3 methyl-butanal and ethanol (55 ppm) produce fruity mushroom/malty aroma and a malty flavor. Moreover, linoleic acid oxidation produces hexanal and methyl ketones yields musty, a metallic, fruity, rancid, and cheese-like aroma and is green in color. Ketones synthesizes 2-nonanone by 2–8 pentanone produces fruity and mushroom-like aromas such as like fruits and mushrooms and green in color. Bacteria synthesizes ketones and chemical fermentation are fortified by mold growth, typically ethyl esters break down from ethanol. Esters produce fruity flavors in fermented sausages (Kosowska et al., 2017).

REFERENCES

1. Abubakr, M. A., Hassan, Z., & Imdakim, M. M. A., (2012). Antioxidant activity of lactic acid bacteria (LAB) fermented skim milk as determined by 1,1-diphenyl-2-picrylhydrazyl (DPPH) and ferrous chelating activity (FCA). African Journal of Microbiology Research, 6(34), 6358–6364.
2. Aka, S., Konan, G., Fokou, G., Dje, K. M., & Bonfoh, B., (2014). Review on African traditional cereal beverages. Am. J. Res. Commun., 2(5), 103–153.
3. Alcántara, C., Bäuerl, C., Revilla-Guarinos, A., Pérez-Martínez, G., Monedero, V., & Zúñiga, M., (2016). Peptide and amino acid metabolism is controlled by an OmpR-family response regulator in Lactobacillus casei. Molecular Microbiology, 100(1), 25–41.
4. Anal, A. K., (2019). Quality ingredients and safety concerns for traditional fermented foods and beverages from Asia: A review. Fermentation, 5(1), 8.
5. Astuti, M. N., Suparmo, M., & Soesatyo, H. N. E., (2013). The role of black soybean tempe in increasing antioxidant enzyme activity and human lymphocyte proliferation in vivo. Int. J. Curr. Microbiol. Appl. Sci., 2, 316–327.
6. Bamidele, O. P., Fasogbon, M. B., Oladiran, D. A., & Akande, E. O., (2015). Nutritional composition of fufu analog flour produced from Cassava root (Manihot esculenta) and Cocoyam (Colocasia esculenta) tuber. Food Science and Nutrition, 3(6), 597–603.
7. Biji, K. B., Ravishankar, C. N., Venkateswarlu, R., Mohan, C. O., & Gopal, T. S., (2016). Biogenic amines in seafood: A review. Journal of Food Science and Technology, 53(5), 2210–2218.
8. Bintsis, T., (2018). Lactic acid bacteria as starter cultures: An update in their metabolism and genetics. AIMS Microbiology, 4(4), 665.
9. Capozzi, V., Fragasso, M., & Bimbo, F., (2021). Microbial resources, fermentation, and reduction of negative externalities in food systems: Patterns toward sustainability and resilience. Fermentation, 7(2), 54.
10. Dajanta, K., Janpum, P., & Leksing, W., (2013). Antioxidant capacities, total phenolics and flavonoids in black and yellow soybeans fermented by Bacillus subtilis: A comparative study of Thai fermented soybeans (Thua nao). International Food Research Journal, 20(6), 3125.

11. De Simone, N., Capozzi, V., Amodio, M. L., Colelli, G., Spano, G., & Russo, P., (2021). Microbial-based biocontrol solutions for fruits and vegetables: Recent insight, patents, and innovative trends. Recent Patents on Food, Nutrition and Agriculture, 12(1), 3–18.
12. Dutra Rosolen, M., Gennari, A., Volpato, G., & De Souza, C. F. V., (2017). Biocatalytic characterization of Aspergillus oryzae β-galactosidase immobilized on functionalized multi-walled carbon nanotubes. Biocatalysis and Biotransformation, 35(4), 260–268.
13. Hao, W., Tian, P., Zheng, M., Wang, H., & Xu, C., (2020). Characteristics of proteolytic microorganisms and their effects on proteolysis in total mixed ration silages of soybean curd residue. Asian-Australasian Journal of Animal Sciences, 33(1), 100.
14. Hasan, M. N., Sultan, M. Z., & Mar-E-Um, M., (2014). Significance of fermented food in nutrition and food science. Journal of Scientific Research, 6(2), 373–386.
15. Hornero-Mendez, D., & Garrido-Fernandez, A., (1997). Rapid high-performance liquid chromatography analysis of biogenic amines in fermented vegetable brines. Journal of Food Protection, 60(4), 414–419.
16. Hou, J. C., Liu, F., Ren, D. X., Han, W. W., & Du, Y. O., (2015). Effect of culturing conditions on the expression of key enzymes in the proteolytic system of Lactobacillus bulgaricus. Journal of Zhejiang University-SCIENCE B, 16(4), 317–326.
17. Juturu, V., & Wu, J. C., (2016). Microbial production of lactic acid: The latest development. Critical Reviews in Biotechnology, 36(6), 967–977.
18. Kalač, P., Špička, J., Křížek, M., & Pelikánová, T., (2000). The effects of lactic acid bacteria inoculants on biogenic amines formation in sauerkraut. Food Chemistry, 70(3), 355–359.
19. Karsloğlu, B., Çiçek, Ü. E., Kolsarici, N., & Candoğan, K., (2014). Lipolytic changes in fermented sausages produced with turkey meat: Effects of starter culture and heat treatment. Korean Journal for Food Science of Animal Resources, 34(1), 40.
20. Kennedy, D. O., (2016). B vitamins and the brain: mechanisms, dose, and efficacy: A review. Nutrients, 8(2), 68.
21. Kirschbaum, J., Rebscher, K., & Brückner, H., (2000). Liquid chromatographic determination of biogenic amines in fermented

foods after derivatization with 3, 5-dinitrobenzoyl chloride. Journal of Chromatography A, 881(1–2), 517–530.

22. Kolesarova, E., (1995). Occurrence and formation of biogenic amines in food. Bull. PV, 34 , 109-122.
23. Kosowska, M., Amajcher, M., & Fortuna, T., (2017). Volatile compounds in meat and meat products. Food Science and Technology, 37, 1–7.
24. Kwak, C. S., Park, S. C., & Song, K. Y., (2012). Doenjang, a fermented soybean paste, decreased visceral fat accumulation and adipocyte size in rats fed with high fat diet more effectively than nonfermented soybeans. Journal of Medicinal Food, 15(1), 1–9.
25. Longhi, D. A., Dalcanton, F., De Aragão, G. M. F., Carciofi, B. A. M., & Laurindo, J. B., (2013). Assessing the prediction ability of different mathematical models for the growth of Lactobacillus plantarum under non-isothermal conditions. Journal of Theoretical Biology, 335, 88–96.
26. Mariod, A. A., Idris, Y. M., Osman, N. M., Mohamed, M. A., Sukrab, A. M., Farag, M. Y., & Matthaus, B., (2016). Three Sudanese sorghum-based fermented foods (kisra, Hulu-mur and abreh): Comparison of proximate, nutritional value, microbiological load, and acrylamide content. Ukrainian Journal of Food Science, (4, Iss. 2), 216–228.
27. Mastanjević, K., Kovačević, D., Frece, J., Markov, K., & Pleadin, J., (2017). The effect of autochthonous starter culture, sugars and temperature on the fermentation of Slavonian Kulen. Food Technology and Biotechnology, 55(1), 67–76.
28. Meerak, J., Yukphan, P., Miyashita, M., Sato, H., Nakagawa, Y., & Tahara, Y., (2008). Phylogeny of γ-polyglutamic acid-producing Bacillus strains isolated from a fermented locust bean product manufactured in West Africa. The Journal of General and Applied Microbiology, 54(3), 159–166.
29. Minh, N. G., (2014). Investigation of pickled water spinach (Ipomoea aquatic) fermentation by Lactobacillus sp. Int. J. Multidiscip. Res. Dev., 1, 71–80.
30. Miyamoto, M., Ueno, H. M., Watanabe, M., Tatsuma, Y., Seto, Y., Miyamoto, T., & Nakajima, H., (2015). Distinctive proteolytic activity of cell envelope proteinase of **Lactobacillus helveticus** isolated from Airag, a traditional Mongolian fermented mare's milk. International Journal of Food Microbiology, 197, 65–71.

31. Montriwong, A., Rodtong, S., & Yongsawatdigul, J., (2015). Detergent-stable salt-activated proteinases from Virgibacillus halodenitrificans SK1-3-7 isolated from fish sauce fermentation. Applied Biochemistry and Biotechnology, 176(2), 505–517.
32. Moslehishad, M., Ehsani, M. R., Salami, M., Mirdamadi, S., Ezzatpanah, H., Naslaji, A. N., & Moosavi-Movahedi, A. A., (2013). The comparative assessment of ACE-inhibitory and antioxidant activities of peptide fractions obtained from fermented camel and bovine milk by Lactobacillus rhamnosus PTCC 1637. International Dairy Journal, 29(2), 82–87.
33. Murata, M., (2005). Membrane interaction of soyasaponins in association with their antioxidation effect. Soy Protein Res., 8, 81–85.
34. Nakajima, N., Nozaki, N., Ishihara, K., Ishikawa, A., & Tsuji, H., (2005). Analysis of isoflavone content in tempeh, a fermented soybean, and preparation of a new isoflavone-enriched tempeh. Journal of Bioscience and Bioengineering, 100(6), 685–687.
35. Nkhata, S. G., Ayua, E., Kamau, E. H., & Shingiro, J. B., (2018). Fermentation and germination improve the nutritional value of cereals and legumes through activation of endogenous enzymes. Food Science and Nutrition, 6(8), 2446–2458.
36. Nout, M. J. R., & Kiers, J. L., (2005). Tempe fermentation, innovation, and functionality: Update into the third millennium. Journal of Applied Microbiology, 98(4), 789–805.
37. Papagianni, M., (2012). Metabolic engineering of lactic acid bacteria for the production of industrially important compounds. Computational and Structural Biotechnology Journal, 3(4), e201210003.
38. Pasini, F., Soglia, F., Petracci, M., Caboni, M. F., Marziali, S., Montanari, C., & Tabanelli, G., (2018). Effect of fermentation with different lactic acid bacteria starter cultures on biogenic amine content and ripening patterns in dry fermented sausages. Nutrients, 10(10), 1497.
39. Paucar-Menacho, L. M., Amaya-Farfán, J., Berhow, M. A., Mandarino, J. M. G., De Mejia, E. G., & Chang, Y. K., (2010). A high-protein soybean cultivar contains lower isoflavones and saponins but higher minerals and bioactive peptides than a low-protein cultivar. Food Chemistry, 120(1), 15–21.
40. Qian, B., Xing, M., Cui, L., Deng, Y., Xu, Y., Huang, M., & Zhang, S., (2011). Antioxidant, antihypertensive, and immunomodulatory

activities of peptide fractions from fermented skim milk with Lactobacillus delbrueckii ssp. bulgaricus LB340. Journal of Dairy Research, 78(1), 72–79.

41. Rollán, G. C., Gerez, C. L., & LeBlanc, J. G., (2019). Lactic fermentation as a strategy to improve the nutritional and functional values of pseudocereals. Frontiers in Nutrition, 6, 98.
42. Sanjukta, S., & Rai, A. K., (2016). Production of bioactive peptides during soybean fermentation and their potential health benefits. Trends in Food Science and Technology, 50, 1–10.
43. Savijoki, K., Ingmer, H., & Varmanen, P., (2006). Proteolytic systems of lactic acid bacteria. Applied Microbiology and Biotechnology, 71(4), 394–406.
44. Sharma, R., Garg, P., Kumar, P., Bhatia, S. K., & Kulshrestha, S., (2020). Microbial fermentation and its role in quality improvement of fermented foods. Fermentation, 6(4), 106.
45. Tamang, J. P., Dewan, S., Tamang, B., Rai, A., Schilinger, U., & Holzapfel, W. H., (2007). Lactic acid bacteria in Hamel and Marcha of North East India. Indian Journal of Microbiology, 47(2), 119–125.
46. Tamang, J. P., Shin, D. H., Jung, S. J., & Chae, S. W., (2016). Functional properties of microorganisms in fermented foods. Frontiers in Microbiology, 7, 578.
47. Tsuyoshi, N., Fudou, R., Yamanaka, S., Kozaki, M., Tamang, N., Thapa, S., & Tamang, J. P., (2005). Identification of yeast strains isolated from Marcha in Sikkim, a microbial starter for amylolytic fermentation. International Journal of Food Microbiology, 99(2), 135–146.
48. Verni, M., Rizzello, C. G., & Coda, R., (2019). Fermentation biotechnology applied to cereal industry by-products: nutritional and functional insights. Frontiers in Nutrition, 6, 42.
49. Vieco-Saiz, N., Belguesmia, Y., Raspoet, R., Auclair, E., Gancel, F., Kempf, I., & Drider, D., (2019). Benefits and inputs from lactic acid bacteria and their bacteriocins as alternatives to antibiotic growth promoters during food-animal production. Frontiers in Microbiology, 10, 57.
50. Xiang, H., Sun-Waterhouse, D., Waterhouse, G. I., Cui, C., & Ruan, Z., (2019). Fermentation-enabled wellness foods: A fresh perspective. Food Science and Human Wellness, 8(3), 203–243.

51. Yoshii, K., Hosomi, K., Sawane, K., & Kunisawa, J., (2019). Metabolism of dietary and microbial vitamin B family in the regulation of host immunity. Frontiers in Nutrition, 6, 48.
52. Zeng, W., Li, W., Shu, L., Yi, J., Chen, G., & Liang, Z., (2013). Non-sterilized fermentative co-production of poly (γ-glutamic acid) and fibrinolytic enzyme by a thermophilic Bacillus subtilis GXA-28. Bioresource Technology, 142, 697–700.
53. Zhang, Y., Hu, P., Lou, L., Zhan, J., Fan, M., Li, D., & Liao, Q., (2017). Antioxidant activities of lactic acid bacteria for quality improvement of fermented sausage. Journal of Food Science, 82(12), 2960–2967.

Chapter

8 Medical Applications of Industrial Fermentation

CONTENTS

8.1. INTRODUCTION

Fermentation, which was once associated with yeast's anaerobic transformation of sugar to alcohol and carbon dioxide (CO_2), is now being used to describe an industrial method for producing a wide range of biomaterials and metabolites by utilizing microorganisms in regulated culture conditions. Based on the item, fermentation can also be done in a batch process, combinatory fed-batch process, and as a continuous process. Antibiotics, solvents like ethanol, or probiotics like yogurt, intermediary compounds like citric acid, and other medically significant goods have all been produced using fermentation mechanisms for a long time. Antiviral drugs, therapeutic recombinant DNA and proteins, and monoclonal antibodies (MAb) are just a few examples of new era fermentation goods. Fermentation is mostly used for the industrial manufacturing of materials; essential for the production of symptom kits, drug delivery systems, and medical equipment, in addition to drugs. This technology is still at the center of today's fast-growing biopharmaceutical sector, which is projected to grow even faster in the coming years as novel, focused drug development progresses.

Microorganisms are illness-generating organisms in the sight of humans. In reality, only a small percentage of the total microbial community is linked to disease fabrication, with the rest being either harmless or useful to humans. Owing to their potential to produce several chemical compounds that are beneficial to us, such tiny microbes, which range in size less than a micrometer to a few micrometers, are often referred to as small chemical plants. Microorganisms manufacture these materials as a result of their metabolic methods, which transform raw materials into simpler molecules to generate energy. To break down complex substances into simple metabolic pathways, microorganisms use the pentose phosphate pathway, the glycolytic pathway, amino acid metabolism pathway, tricarboxylic acid cycle (TCA), fermentative pathway, glyoxalate shunt, and numerous other interconnected pathways. Glycolysis, the TCA pattern, and fermentative paths are the energy-producing routes that propel other cellular methods. The TCA cycle is triggered during aerobic growth, while the fermentative mechanism is activated during anaerobic progress (Campbell, 1993; Babbar and Jain, 2006). Although fermentative processes are anaerobic metabolic procedures, the word-fermentation—is used in a wider micro-biological perspective to relate to all methods (anaerobic, microaerobic, and aerobic) that take to the development of a commodity via mass culture of an entity (Stanbury et al., 1995; Rahman, 2013). Simple organic compounds are the final goods of fermentation, that can be used by the same or by other species. Numerous transitional compounds are emancipated into the growing medium in regards to the final products, that may be valuable to the identical or other microbes. Industries utilize certain microorganisms to start producing particular products of concern in a regulated environment within a container termed as fermenter, due to the significance of these intermediate or final substances for human usages (Figure 8.1).

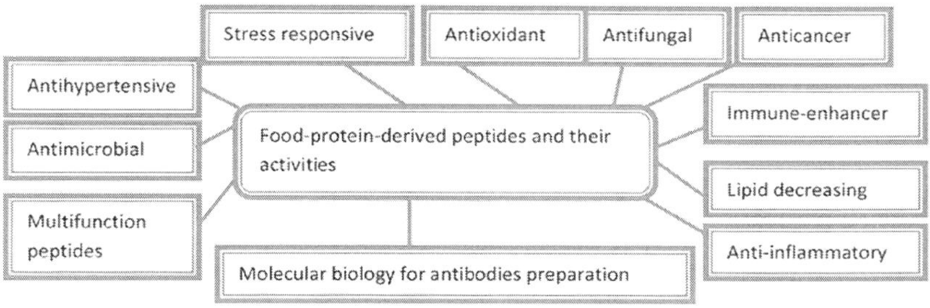

Figure 8.1. Microbial proteases having medical utilization/application.

Source: https://www.frontiersin.org/articles/10.3389/fbioe.2019.00110/full.

Fermentation had been used 1,000 years back by the citizens of Greece, Mesopotamia, the Indian subcontinent, China, Egypt, Rome, and several other regions of the globe in manufacturing conventional homemade beverages, medicine, and foods. Whereas Louis Pasteur described the main principle and theory of fermentation during the second half of the 19th century. Ancient fermentation was primarily used to preserve food. Nevertheless, as early as 6000 BC, the Sumerians, and the Babylonians used yeast to brew beer as social lubricants. It is thought that the Chinese citizens used alcoholic drinks in medication even before that period. They utilized alcohol either like medicine or to disintegrate spices, herbs, and other medically significant organisms. From 3150 Before Christ, alcoholic drinks were used as psychotropic, antimicrobials, and anesthetic medicines in primitive Egypt. They also used a variety of preservatives with wine for cancer therapy and the embalming of dead bodies (Das and Deka, 2012; Bell et al., 2017).

Even though fermentation items were conventionally used to make food, they are now extremely versatile, and many industries that depend on yeast have arisen over the last century to generate a wide range of ingredients that become indispensable in our everyday lives. Biofuels, alcoholic beverages (like wine and beer), food supplements (such as pickles, yogurt, and cheese,) and food, pharmaceutical chemicals (for instance, vaccines, and enzymes), agricultural products (for example, microbial pesticides), industrial chemicals (such as butanol and acetone), and analytical products (like, citric acid) are now manufactured by fermentation industries (Ledingham, 1953). Though accurate proportions of fermentation goods utilized in medical-application are hard to calculate, this is a booming industry with a large variety of items marketed and developed in immediate years.

8.2. FERMENTATION IN INDUSTRY

Industrial fermentation is based on customary principles, yet it encompasses the most significant advances in engineering and biotechnology to increase yield while lowering costs. The organisms chosen, the means used to thrive them, as well as the end products are all important considerations. Furthermore, the size and type of the fermenter whereby the fermentation procedure is carried out are critical considerations. Bio-reactors are generally fabricated of stainless-steel containers, but there are other different types, like those that integrate stainless steel and glass structures, and their holding capacity can range from a few 100 liters to 10,000 liters. The fermenters must be capable of maintaining: (i) a culture surrounding with appropriate nutrient concentrations, temperature, redox potential, dissolved oxygen, and pH; (ii) agitation to make sure that substrates, heat, cells, and air are distributed evenly; (iii) high cell levels for high end-product yield; and (iv) sterile requirements. They have different characteristics depending on the type, but they all have propellers to maintain aeration, a temperature controller and sensor, a computerized system to monitor all parameters, an oxygen detector, the species' growth condition, an inlet to add platform and inoculated, adjustor, and a pH detector, and an outlet to recover products (Rahman, 2013). As shown in Figure 8.2, amongst the most familiar types of reactors used during the fermentation industry is the stirred-tank reactor, which is usually cylindrical in structure and has multiple blades positioned on a shaft operated by a motor. Fed-batch, continuous, and batch fermentations are the three types of industrial fermentation performed out in fermenters. The strategy chosen is determined by the organism's raw material, target product, and type.

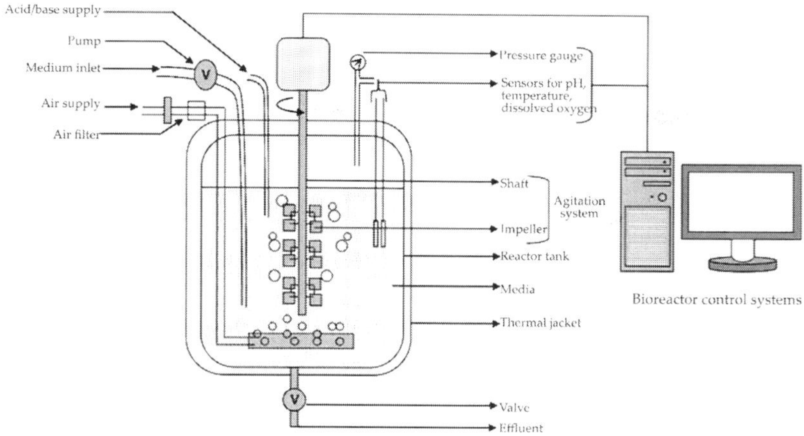

Figure 8.2. A typical bioreactor and its various components are depicted in the diagram.

Source: https://www.slideshare.net/sabanaeem1/bioreactor-and-its-applications.

8.3. INDUSTRIAL FERMENTATION TYPES

8.3.1. Batch Fermentation

In batch fermentation, the growth of a microorganism is via the presence of a fixed number of nutrients and substrates that are inserted at the start of the fermentation process. It is executed in fermenters that can keep substrates in high-working volumes, whereas inoculation is formulated in T-tubes or baffled flasks. In batch fermentation, cell growth follows quintessential microorganism growth patterns: stationary phase, the lag phase, early stationary phase, and exponential phase. This has the obvious disadvantage of low productivity and high turn-around times due to the need to wash and sterilize the fermenters. Batch fermentation, on the other hand, is the primary approach for learning about an important industrial organism's growth parameters, such as CO_2 evolution rate, its specific rate of growth, oxygen consumption, final product yield, and the consumption of substrate rate, and the different types of goods manufactured from the utilization of particular substrates. Batch fermentation is also used by scientists to optimize the operating parameters of the fermentation method, such as pH, aeration rate, temperature, agitation rate, and media volume (Figure 8.3) (Rahman, 2013; Saroj et al., 2014).

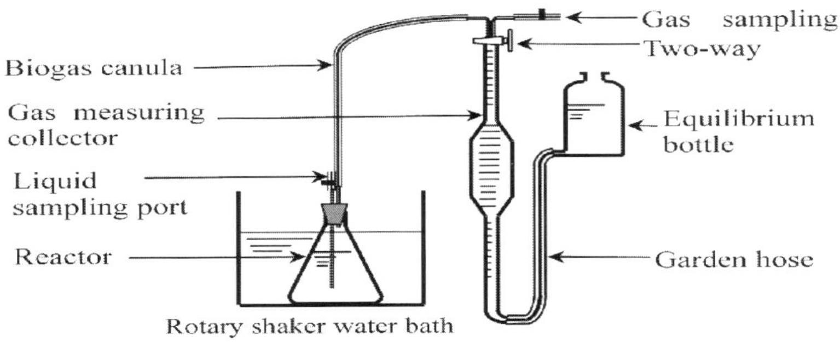

Figure 8.3. The batch fermentation process is depicted schematically in the diagram.

Source: https://www.researchgate.net/figure/Schematic-diagram-of-batch-culture_fig2_224914392.

8.3.2. Continuous Fermentation

The continuous fermentation process involves regulating substrate input and end-product output to maintain a particular growth rate of the microbe. The main benefit of continuous fermentation is high yield, yet because of the high dilution of the final product, it is more susceptible to contamination and product retrieval could be costly (Rahman, 2013). The main difference between continuous and batch fermentation processes is that in batch fermentation, the substrate is just inserted once. On the contrary, in continuous fermentation—the substrate is inserted intermittently with the commodity being withdrawn at about the same time (Figure 8.4).

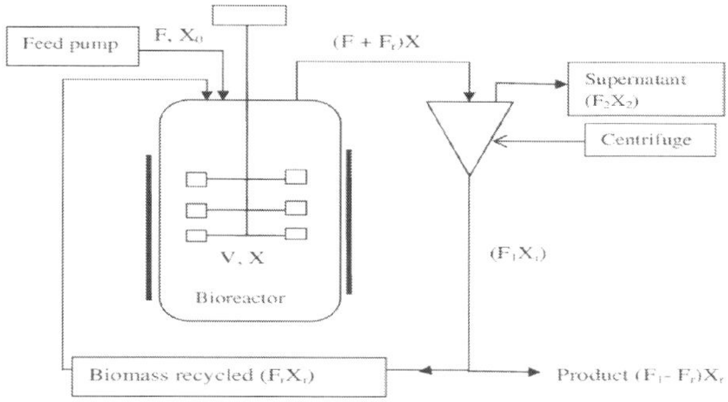

Figure 8.4. Continuous batch fermentation diagram.

Source: https://www.researchgate.net/figure/Schematic-diagram-of-continuous-fermentation-with-cell-recycle-system_fig8_258253836.

Another significant distinction between these two methods is that batch fermentation uses undefined or complex media for yeast growth. Whereas continuous fermentation uses a specified medium (Saroj et al., 2014). Whilst both of these methods are effective at learning about organisms' metabolic processes by combining biochemical assays-proteomics and gene expression assessment-transcriptomics, continuous fermentation allows for a more accurate evaluation of the surfaces used at various parts of the metabolic paths. When the specified media includes ^{13}C labeled substrates like acetic acid (AA) or glucose or other labeled substrates, it is sufficient to analyze the metabolic flux of a particular substrate using NMR-Nuclear Magnetic Resonance techniques and mathematical modeling (Rahman and Shimizu, 2008; Matsuoka and Shimizu, 2011). Metabolomics is the study of an organism's metabolic activities. Scientists use metabolomic, proteomic, and transcriptomic data to genetically modify microbes to create a high rate of productivity by industrial bacteria. Recombinant deoxyribonucleic acid (DNA) technology is used to modify genes, and researchers have recently formed genetically modified yeast, bacteria, viruses, and even vertebrate cell-line for commercial fermentation methods, like CHO-Chinese Hamster Ovarian cells.

8.3.3. Fed-Batch Fermentation

The fed-batch fermentation procedure, which combines both continuous and batch fermentation, is largely used for commercial applications. The organism may be emphasized by the propeller stirring speed, substrates, and pH as well as by the by-products produced throughout the bio-logical transformation of basic materials, through the industrial-scale fermentation procedure. This fermentation process permits the discharge of by-products at levels when they may be hazardous to bacteria, and it does so at intermittent levels. Sub-merged fermentation and solid-state fermentation (SSF) are the two types of fed-batch fermentation. To lower the price of the product recovery phase, both SMF and SSF have been streamlined depending on the type of raw material or substrates used (Subramaniyam and Vimala, 2012).

8.3.3.1. Solid-State Fermentation (SSF)

Solid-state fermentation (SSF) is founded on the concept that solid substances like bran, hay, bagasse, fruit, paper pulp, rice straw and rice, wheat bran, and synthetic media, vegetable waste, and coconut coir are utilized gradually by bacteria at a steady-state rate, resulting in the regulated discharge of nutrients. This method has the advantage of allowing a similar substrate to be utilized for a prolonged time and allowing nutrient-rich waste materials to be reprocessed as fermentation substrates. Conversely, bacteria that need high water activity cannot be used with this method, and purity of the end product is usually expensive (Figure 8.5) (Subramaniyam and Vimala, 2012).

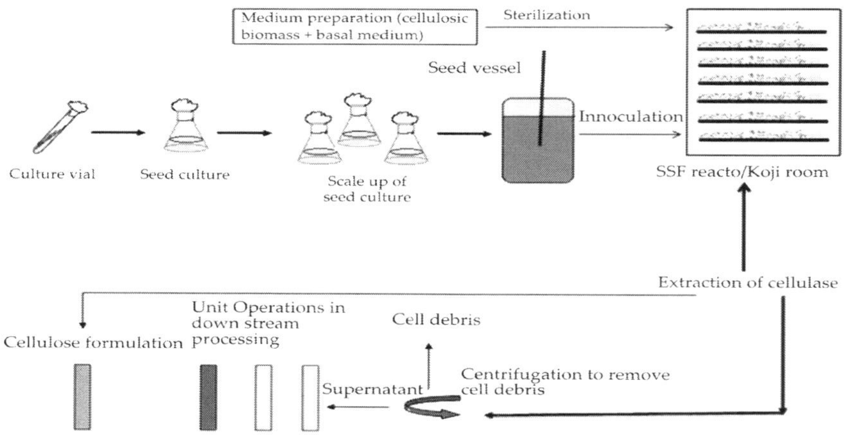

Figure 8.5. An overview of cellulase synthesis using the Koji chamber as solid-state bioreactor of fermentation.

Source: https://www.sciencedirect.com/topics/engineering/solid-state-fermentation.

8.3.3.2. Submerged Fermentation (SMF)

Liquid fermentation (LF)/submerged fermentation (SMF) is a type of fermentation that employs free-flowing fluids like broth and molasses and is ideal for bacteria that demands a lot of water. During this method, the items of interest are discharged into the fermentation broth, making product cleansing simpler. It does, nevertheless, necessitate a steady supply of surfaces. This method is typically used to retrieve secondary metabolites that must be utilized in a liquid state. Soluble sugars, liquid media, wastewater, vegetable juices and fruit, and molasses are all familiar substrates for LF or SMF. Several biotechnology industries which practice E. coli strains to produce commercially essential plasmid DNA use SMF, and they have established proprietary limited media for this reason (Figure 8.6) (Carnes et al., 2011; Subramaniyam and Vimala, 2012).

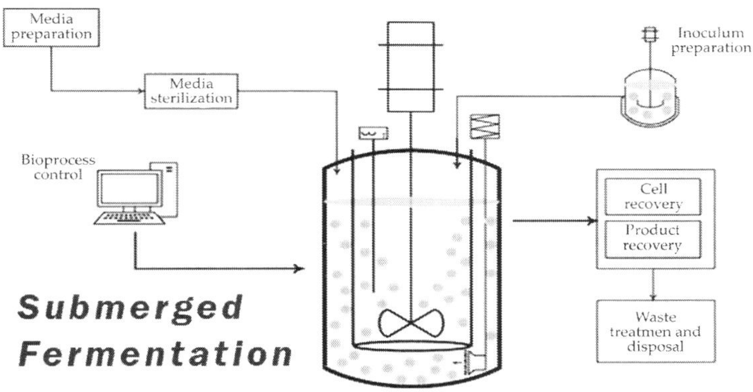

Figure 8.6. Schematic representation of submerged fermentation.

Source: https://microbenotes.com/submerged-fermentation/.

8.4. FERMENTATION STAGES

Down-stream and up-stream methods are terms used by biotechnology industries to describe the phases of the manufacturing procedures. Microorganisms are grown in the bioreactor for product development in the upstream method. To minimize batch-to-batch alteration in production yield, it is analytical to optimize upstream procedures via Research and Development. This is accomplished by identifying the organisms that will produce the highest yield and optimizing their growing conditions in the bioreactor. In the fermentation industry, fungi, and bacteria are the most frequently used microorganisms. Because the fermentation method is entirely dependent on the biochemical reactions carried out by these microbes, they are also known as "bio-catalysts." To generate a significant quantity of products from a specific raw material, the microbe must be able to thrive rapidly and strenuously, be non-pathogenic, and have steady characteristics (Babbar and Jain, 2006). Nowadays, recombinant DNA technology is employed to efficiently produce required products. Mammalian cells, in regards to microorganisms, are inherently modified to produce recombinant proteins (Jayapal et al., 2007). The very first stage in using genetically modified cells is to choose the adequate expression structures to reduce batch-to-batch alteration in product yield. The form of substrate, additionally the organism, is a significant factor in the commercial fermentation procedure. The surface, also referred to as the raw material, must be able to decompose by the bacterium, as well as inexpensive, cost-effective, and easily accessible. A simplified substrate reduces the cost of product recovery in the downstream step.

In contrast to up-stream methods, which are concerned with product development and microbe growth, down-stream methods are concerned with purification and product recovery, product preparation, and product concentration for industrialization as illustrated in Figure 8.7 (Babbar and Jain, 2006). So, when the end-product is not discharged into the substrate, the downstream procedure begins with the fissure of cells by homogenizing, grinding, or using enzymatic compounds. Different kinds of chromatographic separation processes-centrifugation, precipitation, sedimentation, and filtration all are common purifying methods-like ion-exchange chromatography, HPLC (high-performance liquid chromatography), affinity chromatography, and hydrophobic chromatography. After purification, product lines may need to be concentrated using strategies like desiccation, crystallization, lyophilization, evaporation, and solvent extraction (Denyer, 1995).

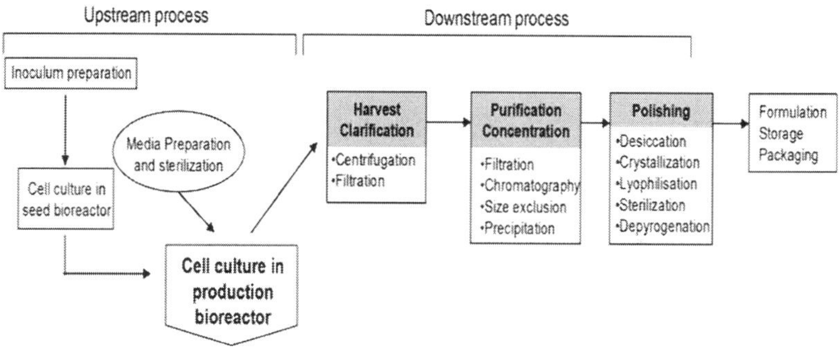

Figure 8.7. The two primary stages of the fermentation process are depicted in the flow diagram.

Source: https://www.scientific.net/AMR.810.127#:~:text=New%20generation%20fermentation%20products%20include,delivery%20vehicles%20and%20medical%20devices.

8.5. FERMENTATION PRODUCTS HAVING MEDICAL SIGNIFICANCE

Conventional medicines made with fermentation methods have been around for 1,000 years. For instance, the expertise of ethanol's pain-killer and disinfectant properties, as well as the conservation and improvement of fermented beverages and foods' nutritional and pharmacological values, was standardized during history (McGovern et al., 2004). Anti-microbials have also been renowned for their action for a long time, as evidenced by the use of moldy soybean cards for the medication of boils and molds for the therapy of foot infectious diseases in the Chinese pharmaceutical industry. Sir Alexander Flemings invented the first antibacterial, penicillin, in 1929, marking the beginning of the antibacterial era. This was accompanied by substantial scientific research to ramp up penicillin development from the lab scale to a commercial scale that was thousands of times larger, establishing the precedent for commercial antibacterial production (Figure 8.8).

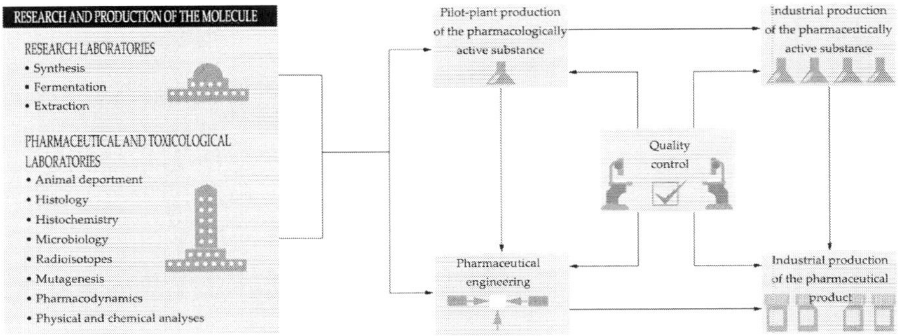

Figure 8.8. The pharmaceutical industry's drug production.

Source: http://www.ilocis.org/documents/chpt79e.htm.

Since 1940, many anti-biotics have been in medical use and have grown substantially due to their capability to kill a wide range of microorganisms and their low toxicity when used properly. With much more understanding of the pathogenicity and strategies of multiple human diseases, fermentation mechanism is now being used to yield a burgeoning number of unmarked, clinically essential products. The complete bacterial cells; enzymes, primary, and secondary metabolites, recombinant bio-molecules, and bio-transformation products, are examples of products (Stanbury et al., 1995; Babbar and Jain, 2006). Fermentation goods are now being utilized in medical trials to treat a wide variety of medical problems, such as immunological disorders, infectious diseases, malnutrition, cancers, metabolic disorders, and cardiovascular diseases.

8.5.1. Antimicrobial Drugs

Antibacterial drugs are compounds generated by definite living organisms at a specific growth rate that constrain other microorganisms. One of the most prominent and frequently used types of fermentation compounds is antimicrobial products. These are antiparasitic, antibacterial, antiviral, and antifungal compounds that are utilized to cure microbial infections in animals and humans. For the industrial production of antibacterial drugs, both liquids submerged and SSFs are employed with the selection procedure based on the type of product yield, substrate, and organism (Song, 1991; Babbar and Jain, 2006). As illustrated in Table 8.1, some of the most commonly used antimicrobial drugs and their mechanisms of action (Figure 8.9).

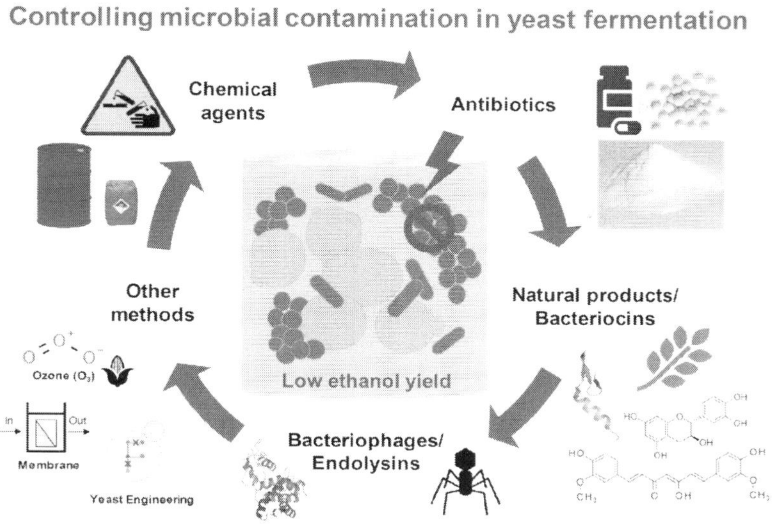

Figure 8.9. In alcoholic fermentation, several strategies for preventing undesired yeast and bacteria contamination.

Source: https://www.mdpi.com/2076-2607/8/2/274/htm

Table 8.1. Fed-Batch Fermentation Method Producing Antimicrobial Substances

Antibiotic	Mode of Action	Source Organism/Antimicrobial Compound
Antibacterial	Cell wall synthesis inhibition	Streptomyces Orientalis/Vancomycin P. griseofulvin/Griseofulvin Penicillium chrysogenum/Penicillin Bacillus subtilis/Bacitracin Cephalosporium acremonium/Cephalosporin
	Cytoplasmic membrane damaged	S. nodosus/Amphotericin Streptomyces noursei/Nystatin Tyrocidins/Bacillus spp., Ploymixin, and gramicidins
	Proteins and nucleic acid synthesis inhibition	Streptomyces/Streptomycin Streptomyces spp. griseus/Tetracyclines Streptomyces venezuelae/Chloramphenicol S. erythaeus/Erythromycin

Antifungal	The fungal cell membrane formation inhibition.	Streptomyces noursei/Polynes
Antiviral	Mitosis inhibition	Penicillium griseofulvin/Griseofulvin
	Viral protein synthesis Inhibition	E. coli, Saccharomyces cerevisiae/Recombinant α-interferon (IFN-α)
	Neuraminidase inhibitor	Shikimic acid (precursor of oseltamivir)
Antiparasitic	Limit the transmission of electrical activity in nematode muscle and neurons cells.	Streptomyces avermitilis/Avermectins Streptomyces sp./Milbemycins

8.5.1.1. Antibacterial

Antibacterial are chemotherapeutic entities derived from microorganisms as secondary metabolites that either inhibit the growth of specific bacteria-bacteriostatic- or trigger the bacteria's killing-bactericidal. Molds naturally produce the majority of antibiotics presently offered in the market. Gram-positive bacteria, on the other hand, emit antibiotic agents that have a bactericidal effect on gram-negative bacteria (living organisms). Penicillin was the first antibacterial agent to be commercialized. Even though Penicillium notatum was utilized in the early stages of manufacturing for this antibacterial drug, and both SMF and SSF were used to relate product yields, the mold Penicillium chrysogenum is now utilized for industrial production by using the SMF procedure. Several other antibacterial drugs are industrially produced utilizing fed-batch fermentation, besides penicillin. Antibacterial drugs are categorized in a variety of ways, most of which are based on their mode of action and structure, like: (i) protein synthesis inhibition and nucleic acid; (ii) inhibition of specific enzyme structures; (iii) inhibition of cell wall synthesis; and (iv) cytoplasmic membrane damage. They are classified as a wider spectrum-influence both Gram-negative and positive bacteria-or narrow-spectrum-influence only Gram-negative or positive bacteria-based on the microbial species they affect (Song, 1991; Denyer, 1995).

8.5.1.2. Antifungal

Antifungals are drugs that particularly kill fungi while causing little harm to the host cells. Antifungal substances are excreted by green-blue algae, bacteria, and molds, while most antimicrobial agents are manufactured by molds. Mycotic infections originated by filamentous yeasts are treated with antibacterial entities like the azole class of drugs. Antifungal medicines are categorized similarly to antibacterial medicines premised on how they work, like: (i) membrane inhibition of DNA fabrication; (ii) inhibition of fungal-cell-wall fabrication; and (iii) disruption of fungal cellular. Antifungal drugs are commercially produced using both fermentation and synthetic methodologies (Biondi et al., 2004; Abdulla and Saeed, 2015).

8.5.1.3. Antiviral

Viral microbes (infections) are treated with antiviral medicines. Antiviral medicines are difficult to make because they must be supplied within the host cell and constrain viral protein manufacturing, unlike antiparasitic, antibacterial, and antifungal drugs, which attack the microorganism explicitly outside of the host cell. As a result, antiviral medications may also be noxious to host-cell. Synthesizing the right antiviral medicines necessitates a thorough understanding of the virus's genetics as well as the genetic factors of the host-cell units, so that the medicine may have the least harmful impact on the host-cell units. Antiviral medications are prescribed to attack the viral protein synthesis system and attachment, and there should be a wide spectrum to be useful against a large number of infections in a similar family or group. Amantadine, interferon, acyclovir, and Tamiflu, are some well-known antiviral vaccines (Soomro et al., 2012). The host defense glycoprotein-interferon, which was first formulated from human WBC (white blood cells) for medical application, is one of these medicines. It is presently being manufactured at a mass-production scale in E. coli and yeast utilizing recombinant technologies and fermentation technologies due to high production costs and scarcity (Shiloach and Fass, 2005; Tian et al., 2017). Tamiflu, a broadly employed oral antiviral medicine for the prevention and treatment of seasonal flu, is made from shikimic acid, a fermented product, using complex azide chemical composition (Li et al., 1999; Ghosh et al., 2012).

8.5.2. Antiparasitic Drugs

Distinctive parasites have multiple ways of interfering with the innate immunity of the host cell. Malaria parasites-Plasmodium sp.-, dengue fever parasites, and other parasites all act as infectious diseases vectors. A few more parasites, on the contrary, are directly linked to diseases, like "Trichomonas vaginitis"-STD (sexually transmitted disease). Nitrifurimox, metronidazole, melarsoporal, and suramin are some familier anti-protozoal medicines. Streptomyces sp.-fermentation produces macrolide anti-biotics with effective antihelminthic activity, like milbemycins and avermectins (Zhinan and Peilin, 1999).

8.5.3. Antitumor Antibiotics

Several anti-biotics cited as antitumor antibacterial drugs are also employed to cure tumors in regards to these antimicrobial entities. Streptomyces types produce anthramycins, which belong to the anthramycin family. Anti-tumor antimicrobial agents are cytotoxic to healthy cells because they attack the function and structure of the genetic code. Many new anti-tumor antimicrobial drugs, on the other hand, are configured to have a lower impact on healthy cells it has been used to medicate breast cancer, sarcomas, acute leukemia, and a variety of many more cancers (Rappa et al., 1993).

8.5.4. Steroids

Steroids are a type of organic compound with cycloalkane rings that can be found in both plants and animals. They have a wide range of clinical applications. These can be employed as anti-cancerous and anti-inflammatory medicines, as well as for improving family planning, asthma immune system, pain treatment, autoimmune disease treatment, anemia treatment, and organ transplantation. The chemical structure of steroids is time-consuming and costly. Animal or plant-derived steroids, on the contrary, can be bio-transformed into therapeutically useful medicines by particular species of fungi, bacteria, and Actinomycetes. For instance, Corynebacterium simplex converts hydrocortisone to prednisolone, Streptomyces sp. converts 9-fluoro hydrocortisone to 9-fluoro-16-hydroxy hydrocortisone, a hormonal product; and an anti-inflammatory drug; Rhizopus sp. converts progesterone to 11-hydroxy-progesterone, a hormonal product. SMF is now used in the commercial production of steroid building block substrates, which are incorporated towards the end of the growing phase for biotransformation (Kapoor et al., 1983; Khan et al., 2012).

8.5.5. Hypocholesterolemic Agents

Hypocholesterolemic entities are compounds that prevent the liver from producing too much cholesterol. Elevated blood cholesterol levels-"hypercholesterolemia"-arethoughttobethemajorreasonsforatherosclerosis, which leads to cardiovascular disease, so these entities are extremely important in medicine. Mevastatin, compactin, simvastatin, lovastatin, and pravastatin are amongst the most frequently used hypocholesterolemic agents. Statins are the names of the last four substances. The other three medicines are developed by Streptomyces carbophilus, Aspergillus terreus, and Penicillium citrinum, except "simvastatin"-which is semi-synthetically obtained from lovastatin (Manzoni and Rollini, 2002). Statins are not exclusively used as hypo-cholesterolemic entities. They are also being used to cure a variety of diseases, including bone disorders, Alzheimer's disease, renal protection, multiple sclerosis, and acute lymphoid leukemia. Lungs, colorectal, prostate, pancreatic cancers, and breast have all been linked to statin therapy.

Both SMF and SSF-solid state have been utilized for large-scale production of statin and similar compounds; after all, SSF is chosen due to product stability, increased porosity, increased product yield, and lower media prices (Praveen and Savitha, 2012).

8.5.6. Antihypertensive Agents

Anti-hypertensive proteins are a type of peptide that are used to stop the ACE-angiotensin-converting enzyme from working. Hypertension is mainly due to high blood pressure that is regulated by ACE. ACEIA-"angiotensin-converting enzyme inhibitory agents"-are inhibiting peptides that are suggested to be incorporated in a human's daily meals to minimize stress levels. Several fermented foods, including peptide, supplemented fermented soy-bean flavorings, enzyme improved cheeses, and fermented dairy products encompass anti-hypertensive substances. A few conventional fermented products, like nyufu, natto, and douchi, also comprise bio-active substances with ACEIA. For the manufacturing of these goods, both liquid and SSF are employed, with Lactobacillus organisms being the most frequently utilized bacteria (Subramaniyam and Vimala, 2014).

8.5.7. Alkaloids

Alkaloids are a class of chemical substances that are manufactured by a wide range of microorganisms, including fungi, animals, bacteria, and plants, contain primarily fundamental N_2 atoms. Secondary metabolites are substances that have a long record of use as conventional drugs. Several alkaloids have antiviral, antimicrobial, and anti-carcinogenic properties. Chemical fabrication methods are still used for the industrial production of alkaloids, but due to small yields and high costs, fermentation methodologies are being considered for bulk manufacturing. The SMF procedure for, Trichoderma harzianum, and the fungus Myrothecium verrucaria with ethyl and methyl alcohol as raw materials are being researched (Subramaniyam and Vimala, 2014).

8.5.8. Solvents

In medicine, alcohol is a frequent solvent that is being used. It is being used in things like anti-depressant medicines, cough syrups, and tincture iodine. Ethanol is the main metabolite created by fermentable carbohydrates such as grapes, maize starch, sugar beets, molasses, and potatoes, which are converted into CO_2 and ethanol by the yeast Saccharomyces cerevisiae. The predominant fermentation technology for industrial ethanol production is SSF (Kapoor et al., 1983). In medicine, glycerol is another extensively used solvent. It is a basic sugar alcohol that is also termed glycerin or glycerine. In elixirs and tinctures like Theophylline, which is utilized to cure asthma, glycerol is employed as a solvent. Also, it is utilized as a carrier in several drugs for the healing of eye problems, constipation, cerebral edema, and vasodilators. Glycerol is a preservative that is used in lotions and ointments to keep them against drying out. Chemical synthesis or microbial fermentation can both produce glycerol. But microbial fermentation is the most extensively employed technique since it is a byproduct of the yeast Saccharomyces cerevisiae sugar to ethanol fermentation, and strains of S. cerevisiae that is genetically engineered are now being used to boost product production (Wang et al., 2001).

8.5.9. Intermediate Compounds

Citric acid is indeed a metabolic intermediary that can be found in small quantities in practically all animals and plants. Kidney stones, gout, and metabolic acidosis in individuals with renal difficulties are treated with a combination of potassium citrate, citric acid, and sodium. Since many yeast species potentially make citric acid through carbohydrate metabolism, Aspergillus niger yeast produces around 70% of commercially made citric acid, as well as the raw materials utilize, which are sugars, cane molasses, and fruit pulp. This is accomplished through SMF (Kapoor et al., 1983; Papagianni, 2007).

Lactic acid (LA) is a significant source of energy in organisms during the metabolism of carbohydrates. LA has been shown to protect the gray matter of the brain in recent investigations. Even though muscle cramps are a prominent drawback of LA generation, LA derivatives are utilized to cure anemia and calcium ion deficit. LA is produced industrially using the bacterium Lactobacillus bulgaricus with a variety of carbohydrate compounds like whey, potato starch, and molasses. For optimal LA generation, the bacteria require salt, nitrogen, and vitamins including to carbohydrates. As a result, whey is the most commonly utilized raw material in the manufacture of LA (Kapoor et al., 1983).

Fumaric acid is an energy source for humans and other creatures. It is an intermediate component in the cycle of TCA. Fumaric acid esters are currently approved for the cure of multiple psoriasis and sclerosis. Fumaric acid is made commercially utilizing a petroleum raw material called butane. Furthermore, as butane is becoming more expensive therefore fermentation technologies for fumaric acid synthesis are becoming more popular. The main benefit of this method is that glucose is 3-times lower in cost than butane as raw material and the filamentous fungus Rhizopus sp. can create 85% w/w fumaric acid using glucose. For the industrial manufacturing process of fumaric acid, SMF is being utilized, and the product yield and recovery are appealing. To boost product yield, metabolically modified species are being used (Engel et al., 2008; Gold et al., 2012).

Succinic acid is a cycle of TCA intermediate made up of malic and fumaric acid, as well as an anaerobic fermentation final product. Capsules of succinic acid are utilized to cure reperfusion and cardiac ischemia due to a mitochondrial complex I malfunction. Industrial manufacturing of succinic acid-like fumaric acid was reliant on petroleum-based raw materials. But, due to the rising cost of petroleum raw materials, microbial fermentation with renewable raw material as a feedstock has sparked a lot of attention. Succinic acid is made by a variety of bacteria, however, Actinobacillus succinogenes has been recognized as among the highest promising microorganisms for commercial production. Straw, cane molasses, maize cob, crop stalk wastes, sake lee, and other raw materials(bio base) have recently been utilized in commercial fermentation processes (Suno and Nagaoka, 1989).

Pyruvic acid is a significant component of the cycle of TCA since it is a carbohydrate metabolism end product and a crucial substrate for the cycle of TCA, also known as the fermentative route, that generates energy. Pyruvate is utilized in medication to cure heart ischemia, cholesterol, high blood pressure, and obesity. The chemical process is used in the industrial synthesis of this organic acid. Furthermore, because of the high expense of recovering the product, the fermentation procedure for large-scale production is being reconsidered.

While many yeast and bacteria can collect pyruvate, Torulopsis glabrata is among the most prolific strains producing a high yield pyruvate and using glucose as the raw material in the process of fermentation (Li et al., 2001).

Lysine is indeed an important amino acid that is synthesized primarily as an intermediate metabolite during the metabolism of nitrogenous compounds. Cold sores caused by the herpes simplex labialis virus are treated with L-lysine. This can be administered orally or applied directly to the skin sores. Steadily growing E. coli bacteria inside a medium containing ammonium hydrogen phosphate, glycerol, and corn steep-liquor under-regulated pH, aeration, and temperature results in the industrial production of L-lysine. Diaminopimelic acid (DAPA) is the result of this process, which is destroyed by decarboxylase enzymes to create L-Lysine (Kapoor et al., 1983).

Glutamine is one of the most abundant amino acids within the body. It comes mostly from skeletal muscles and is rapidly utilized by the immune system, liver, gastrointestinal tract (GI), and kidneys. It is indeed a crucial product in the formation of many complex carbohydrates, amino acids, pyrimidines, purines, and co-enzymes, and also among the most essential substances in the metabolism of nitrogen. L-glutamine is useful in both the industry and medical fields. It is being used to cure mucositis in radiation therapy and chemotherapy patients. It was also utilized to promote liver function but also as an amino acid infusion. L-glutamine is commercially generated employing Corynebacterium sp. that has been genetically engineered (Akiyama et al., 2003).

Glutamic acid is indeed an unnecessary amino acid that is created from various amino acids like arginine and ornithine, as well as generated as an intermediate result of the TCA cycle's alpha-ketoglutarate metabolism. Glutamic acid, a neurotransmitter, is essential for fat and glucose metabolism in both animals and humans. Glutamic acid is found in specific diets for patients having metabolic disorders, as well as medications that dissolve lungs mucus. Glutamic acid is produced by a variety of fungal and bacterial species. Industrial production is based on fermentation, which employs a variety of Brevibacterium, Micrococcus, and Arthrobacter species (Akiyama et al., 2003).

8.5.10. Enzymes

Due to the medical relevance of several fungal and bacterial enzymes, these are generated commercially using the technology of fermentation. These enzymes are made using both solid-state and SMF. Antitumor effects of enzymes including L-asparaginase and L-glutaminase have gotten a lot of interest in medicine. In the treatment of chronic lymphocytic leukemia, these enzymes are employed in the group. Streptomyces sp. is used to make commercially available L-asparaginase. Pseudomonas aeruginosa sp. is used to make L-glutaminase. Much of those enzymes are employed to decompose basic materials like complex carbohydrates like cellulose and starch to facilitate the process of fermentation for many other medically relevant chemicals. Invertase, esterase, Cellulase, lipase, and some other fungal enzymes are extensively utilized in industrial fermentation. Aspergillus niger is perhaps the commonly employed fungus species for such function (Kapoor et al., 1983).

8.5.11. Recombinant Proteins

Recombinant proteins were currently widely used to cure a variety of disorders in humans, including infertility and cancer. For example, luteinizing hormone is used to treat infertility, an activator of tissue plasminogen (engineered) is used to treat myocardial infarction, TNFα receptor proteins are used to treat rheumatoid arthritis, and a wide range of MAb toward cancer is addressed subsequently in this chapter. A few of the recombinant proteins which are currently being used in clinical trials were shown in Table 8.2. Recombinant proteins are produced using both eukaryotic and prokaryotic methods (Jayapal et al., 2007). From the production to the application stage, there are various stages to consider. This necessitates the use of a genetically modified organism, as well as effective fermentation procedures and delivery methods to the cells of the host. E. coli has also been widely employed to produce recombinant proteins throughout the last many decades. Furthermore, in addition to this bacterium, various mammalian, viral, yeast, and fungal systems for recombinant protein synthesis have recently been created.

Biotechnology and pharmaceutical businesses are increasingly turning to mammalian systems for the production of protein as genomes and proteomics improve when the bacterial system did not contain the necessary machinery to manufacture suitable eukaryotic proteins. It indicates that post-translational alterations and correct folding, like glycosylations, are essential for a protein molecule to be functional and exhibit its complete activity once given to target cells. Because the system of prokaryotes lacks this, therefore many recombinant proteins utilized as medicines are produced in the culture of the mammalian cell such as CHO. CHO cells are currently used to make about 70% of all recombinant therapeutics. Even though protein productivity in the mammalian system was initially low, production capacity has recently enhanced due to macroscopic and microscopic improvements in mammalian culture, including the development of highly amplifiable and selectable expression systems, as well as the isolation of highly expressing and stable clones, which seem to be the pinpoint for enhanced productivity. Fermentation techniques, process management, and media composition were all improvised at the macro level. This subsequent technique has allowed productivity to approach gram per liter levels, with a 100-fold increase in product yield above titers reported in the mid-1980s (Palomares et al., 2004; Carnes et al., 2006).

Table 8.2. A Collection of Recombinant Proteins That Have been Employed for Medicinal Purposes (Liras, 2008)

SL. No.	Brand Name	Type	Therapeutic Use	Company
1.	Epogen	Erythropoietin	Anemia	Amgen
2.	Activase	Tissue plasminogen activator	Acute myocardial infraction	Genentech
3.	Follistim	Follicle-stimulating hormone	Infertility	Merck
4.	ReFacto	Factor VIII	Hemophilia A	Wyeth
5.	Pulmozyme	Deoxyribonuclease I	Cystic fibrosis	Genentech
6.	Humulin	Insulin	Diabetes	Lilly USA

8.5.12. Gene Therapy

Genetic engineering, by definition, is the delivery of nucleotides like medicine to change the direction of a medical disease or condition (Wu and Ataai, 2000). Gene therapy was utilized to cure single gene illnesses like sickle cell anemia, cystic fibrosis, glycogen storage disease, color blindness, blue boy, or severe combined immune deficiency (ADA-SCID) disease, hemophilia, chronic granulomatous disorder (CGD), muscular dystrophy, lysosomal storage disease, and congenital blindness for the past two decades. Several gene therapy procedures have been established to cure a broad range of tumors employing oncolytic virotherapy, suicide gene therapy, therapeutic gene, and anti-angiogenesis vaccines to cure acute disorders. Approximately 2/3 trials of gene therapy are for cancer disease, and most are already in advanced stages of treatment for neck and head cancer, pancreatic cancer, and prostate cancer. Kidney, skin, brain, colon, liver, and breast cancer therapies are also being tested in clinical studies. Parkinson's disease gene therapy and Huntington's disease, in addition to cancer, are currently under Phase I clinical studies. Gene treatments for viral diseases (i.e., hepatitis, influenza, HIV), diabetes, and heart disease are also being researched and will soon be tested in the initial phase of clinical studies (Coulon et al., 2006). Various methods, such as the utilization of liposomes and viral vectors in different applications, have been used to administer therapeutic RNA or DNA. Other non-viral biological entities like bacteriophage, virus-like particles (VLPs), and bacteria are also being studied because of the limits and hazards connected with all of these vehicles, like unfavorable immunological consequences, the complexity of production, and packaging issues (Seow and Wood, 2009). Given these types of gene delivering technologies, fermentation technology is likely to play a crucial role in commercializing these vehicles shortly. For gene therapy, it is also fascinating to learn that research is being done on commercial plasmid DNA synthesis (Lahijani et al., 1996).

8.5.13. Monoclonal Antibodies (MAb)

Development in our understanding of biomolecular disease mechanisms, particularly in immunological disorders, cancers, and multiple sclerosis, have led to the advancement of particular target therapies relying on antibodies of monoclonal that behave specifically on numerous extracellular variables engaged in disease pathogenicity. Bevacizumab or Avastin, a monoclonal antibody that is humanized against the growth factor of vascular endothelial, is being used to cure colorectal cancer in the second line. In secondary breast cancer, Herceptin (Genentech) affects human epidermal growing element receptor 2, while Cimzia affects tumor necrosis factor in Crohn's disorder. The affinity, bioavailability, and monoclonal antibody medicines specificity, but also whether or not the treatments have immunogenic effects, are all important considerations in achieving a successful therapeutic outcome (Mehren et al., 2003; Shukla and Thömmes, 2010). Table 8.3 lists some of the MAb which are employed in medicine.

Table 8.3. A Collection of Monoclonal Antibodies That Have been Utilized for Medical Reasons (Wu and Ataai, 2000)

SL. No.	Type	Brand Name	Therapeutic Use	Company
1.	Anti-IgE mAb	Xolair	Severe asthma/moderate	Genentech
2.	Anti-VEGF mAb	Avastin	Metastatic colorectal and lung cancer	Genentech
3.	–	Humira	Rheumatoid arthritis	Abbott
4.	Anti-CD11a mAb	Raptiva	Chronic psoriasis	Genentech
5.	–	Rituxan	Non-Hodgkin's lymphoma	Genentech
6.	Anti-EGFR mAb	Vectibix	Metastatic colorectal cancer	Amgen

MAb are traditionally made by utilizing the technology of mouse hybridoma, in which mice are implanted with both the antigen and their sera are collected regularly to screen for particular antibody development. After that, these mice are killed and their spleens are extracted. Just the fused cells persist in culture when mice splenocytes are cultured with preconditioned myeloma cells in hypoxanthine-aminopterin thymidine media. The appropriate clone for the productions of antibodies against the targeted antigen is subsequently chosen through a colony screening approach. System of mammalian expressions like Chinese hamster ovary (CHO) cells, NS0 murine myeloma cells, and PER.C6 human cells are now utilized in the commercial development of antibodies. Upon being transfected, using plasmids, containing the antibody heavy and light chain genes, cell lines are screened for a greater production of antibody, stable antibody production, and appropriateness for usage in huge bioreactors. Both the downstream antibody purification procedure and upstream culture cell method have seen considerable advances. For huge cell cultures using bioreactors, suspension cell culture is the chosen method. Until being transported to the manufacturing bioreactor, cells are progressively increased in a sequence of seed and inoculums bioreactors. Centrifugation separates cell detritus and cells from the culture medium carrying the antibody after growth. The capture of antibodies utilizing chromatography of protein A, as well as hydrophobic chromatographic and ion exchange methods, are common downstream operations (Li et al., 2010; Rodrigues et al., 2010).

8.5.14. Vaccines

Vaccines are suspensions of living, killed, attenuated microbes or organism particular elements that induce immunity in humans against a certain disease. Vaccinations are generally employed for preventative purposes, that is, to prevent the individual from a specific disease by a certain organism. So, vaccines can currently prevent a vast number of infectious diseases.

The following are some examples of viral vaccines and inactivated bacteria or live attenuated bacteria: Bacillus Calmette-Guérin (BCG), Smallpox, pertussis, mumps, measles, rubella, polio, Japanese encephalitis, influenza viruses, hepatitis A and whole-cell pertussis. Diphtheria, Tetanus, and anthrax vaccines were also created using pure protein antigens. Vaccines for bacterial meningitis and pneumonia, hepatitis B, and the recently identified vaccines of human papillomavirus (HPV) were developed using newer technology that included conjugation of protein to capsular polysaccharides and ways to manufacture recombinant DNA (Josefsberg and Buckland, 2012; Koff et al., 2013). A few of these vaccinations are also effective in the protection of cancers, like liver or cervical cancer, which are both produced by immunization and oncoviruses, against both hepatitis B and HPV virus, correspondingly, may protect us from these cancers.

The kind of microbiological portion or vaccine to be employed for immunization objectives determines the commercial manufacturing of vaccines. The main goal of today's vaccine manufacturing process is to generate vaccinations that are both more effective and safer. Fermentation has long been an important aspect of the commercial production of bacterial vaccines with both continuous and batch fermentation methods. In addition to animal cells' viral vectors, the virus that resembles particles in insect cells or yeast, and vaccines of DNA generated as plasmid DNA through E. coli, the technology of industrial fermentation has previously been used for the manufacturing of viral vectors in living organisms. The plasmid inoculation encoding a gene for a specific antigen protein, which is then produced by a person's cellular machinery getting the vaccine, results in a long-lasting immunological response. Multiple sclerosis, HIV, and cancer vaccines formulations are all in the process. In DNA vaccine preparation fed-batch fermentation with specified media is used. Fermentation of fed-batch is utilized to produce virus resembling to particles within yeast cells, the yeast Saccharomyces cerevisiae serving as the expression system. The yeast system is also used to make HPV vaccinations and recombinant hepatitis B vaccines (Josefsberg and Buckland, 2012).

8.5.15. Site-Specific Immunotherapies (SSI)

These are also known as site-specific immunomodulators and are being developed to target the patient's immune cells within the tumor microenvironment, to stimulate them to attack the tumor. Many SSI medicines are now in clinical studies to address site-specific disorders including autoimmune or cancer diseases. Every SSI contains inactive microbial elements from a particular species of bacteria recognized to cause illness in a particular organ or tissue such as the skin, lungs, or intestines. The autoimmune condition or cancer type determines the selection of SSI treatment. Selecting the phase of growth during which the protein is created, batch fermentation is being used to manufacture inactive components of bacteria (e.g., LPS) (Wang and Walfield, 2005).

8.5.16. Probiotics

A human digestive tract functions as a miniature microbial factory, with hundreds of parasites, bacteria, and yeast working together to disintegrate complex carbohydrates into simpler ones. The eventual purpose of this operation is to generate energy. That bacteria that are good for human health are referred to as probiotics. They are advised for use as a dietary supplement in diseases that alter the natural gut microflora's functioning like immunosuppressive drug utilized or antibiotic treatment, infectious diarrhea, when the immune response needs to be boosted, or other inflammatory bowel illnesses. Furthermore, probiotics are being studied for the diagnosis of stomach ulcers (particularly induced by H. pylori), infections of the urinary tract, sinus, vaginal, as well as other infections of the respiratory tract, allergies, dental disease, colon cancer, and liver illness. Probiotics by the shape of something like a stool transplant or a combination of pure cultures of intestinal bacterial generated from healthier donors are now being tested in clinical trials to cure an infection of recurrent C. difficile (Heller, 2001; Rohlke and Stollman, 2012).

Probiotics come in a variety of forms, including capsules, powder, and live microorganisms in fermented milk products (e.g., cottage cheese, fermented drinks, yogurt, etc.). Bifidobacterium sp. and Lactobacillus sp. are the two most common bacteria found in probiotics. Probiotics contain yeast and various bacilli. Those bacteria are appropriate for usage as probiotics due to their synergistic properties and fermentative behavior. The organism's growth phase (e.g., static growth phase or logarithmic), product temperature, the composition of chemicals (e.g., nitrogen, acidity, mineral content, carbon, oxygen, and water availability), and probiotics probable interaction with the culture media are all factors that affect probiotics commercial production. Though suspended culture and batch fermentation have traditionally been used to generate probiotic organisms, it has been claimed that continued culture could give both significant productivity and cell yield (Heller, 2001).

8.5.17. Diagnostic Products

Antibodies, enzymes, and recombinant proteins created by fermentation technology can be utilized to manufacture and commercialize diagnostic kits for speedy disease diagnosis, in addition to therapeutic purposes. As an instance, batch fermentation recombinant Plasmodium falciparum HRP-II proteins can be utilized to construct malaria testing kits depending on immunochromatographic or ELISA techniques (Singh et al., 2013). Recombinant Brucella melitensis outermost protein base membrane was synthesized in E. coli and generated via batch fermentation, with potential utility in the serological surveillance of brucellosis induced by Brucella bacteria (Bs et al., 2013). Furthermore, widely used enzymes of diagnostic marker such as alkaline phosphatase, horseradish peroxidase, and luciferase can be synthesized as conjugates containing antibodies utilizing the recombinant DNA technology, accompanied by fermentation, for improved efficiency (Rahman, 2013). The technology of fermentation is particularly significant for the synthesis of recombinant enzymes like Taq DNA polymerase generated by genetically altered E. coli or Thermus aquaticus, which are utilized in several in vitro diagnostic assays based on the nucleic acid. A33scFv, a humanized antibody that targets a glycoprotein of a cell surface-expressed by colon cancer, was developed in Pichia pastoris and generated through fermentation of fed-batch for colon cancer immunotherapy, although it has also been proposed for possible diagnostic applications (Damasceno et al., 2004). Besides these, fermentation has previously been utilized to produce materials for tissue culture and bacteriological media, like sugars, yeast extract, and a variety of other substances.

8.5.18. Biopolymers

In certain dietary circumstances, bacteria synthesize a range of polymeric compounds from diverse intermediates of core metabolism. Polyphosphates, Polysaccharides, polyamides, and polyesters are the four main types of biopolymers. Due to their non-toxicity and biodegradability, biopolymers are frequently used in medicinal purposes. Drug delivery, Tissue engineering, and wound dressing processes have all utilized biopolymers as matrices. Many biopolymers are now being manufactured on a big scale in industrial fermenters. In some situations, metabolic paths of producing species were altered for tailored manufacturing of specified products to increase the applied effectiveness of biopolymers. Alginate, that is manufactured by Azotobacter and Pseudomonas spp., is utilized for drug delivery and tissue scaffolding; dextran, that is developed by Streptococcus and Leuconostoc spp., is utilized to create an extender for blood plasma; and gellan, that is manufactured by Sphingomonas sp., is utilized as a thickening agent and for encapsulation of drug (Rehm, 2010).

REFERENCES

1. Abdulla, K. A., & Saeed, R. R., (2015). What is deep tropical mycosis? An underestimated entity with serious problems. International Journal of Advances in Medicine, 2(2), 76.
2. Abougergi, M. S., & Kwon, J. H., (2011). Intravenous immunoglobulin for the treatment of Clostridium difficile infection: A review. Digestive Diseases and Sciences, 56(1), 19–26.
3. Akiyama, T., Bouillet, P., Miyazaki, T., Kadono, Y., Chikuda, H., Chung, U. I., & Tanaka, S., (2003). Regulation of osteoclast apoptosis by ubiquitylation of proapoptotic BH3-only Bcl-2 family member Bim. The EMBO Journal, 22(24), 6653–6664.
4. Babbar, S. B., & Jain, R., (2006). Xanthan gum: An economical partial substitute for agar in microbial culture media. Current Microbiology, 52(4), 287–292.
5. Bell, V., Ferrão, J., & Fernandes, T., (2017). Nutritional guidelines and fermented food frameworks. Foods, 6(8), 65.
6. Biondi, N., Piccardi, R., Margheri, M. C., Rodolfi, L., Smith, G. D., & Tredici, M. R., (2004). Evaluation of Nostoc strain ATCC 53789 as a potential source of natural pesticides. Applied and Environmental Microbiology, 70(6), 3313–3320.
7. Bs, K., Tn, A., Ashu, K., Tiwari, S., & Singh, K. A., (2013). Fed-batch fermentation and purification strategy for high yield production of Brucella melitensis recombinant Omp 28 kDa protein and its application in disease diagnosis. Protein and Peptide Letters, 20(7), 731–740.
8. Byzova, N. A., Sviridov, V. V., Gavrilova, N. F., Raspopova, E. N., Jakovleva, I. V., Generalova, A. N., & Dzantiev, B. B., (2009). Immunochromatographic and latex-agglutination systems for diphtheria toxin detection. Russian Journal of Bioorganic Chemistry, 35(4), 482–489.
9. Campbell, P. N., (1993). Principles of biochemistry second edition, Worth, New York, 87901, 50-400.
10. Carnes, A. E., Hodgson, C. P., & Williams, J. A., (2006). Inducible Escherichia coli fermentation for increased plasmid DNA production. Biotechnology and Applied Biochemistry, 45(3), 155–166.
11. Carnes, A. E., Luke, J. M., Vincent, J. M., Schukar, A., Anderson, S., Hodgson, C. P., & Williams, J. A., (2011). Plasmid DNA fermentation

strain and process-specific effects on vector yield, quality, and transgene expression. Biotechnology and Bioengineering, 108(2), 354–363.

12. Coulon, J., Husnik, J. I., Inglis, D. L., Merwe, G. K., & Van Der, L. A., Erasmus, D. J., & Vuuren, H. J. J. V., (2006). Metabolic engineering of Saccharomyces cerevisiae to minimize the production of ethyl carbamate in wine. American Journal of Enology and Viticulture, 57, 113–124.
13. D'Argenio, D. A., & Wilson, C. B., (2010). A decade of vaccines: Integrating immunology and vaccinology for rational vaccine design. Immunity, 33(4), 437–440.
14. Damasceno, L. M., Anderson, K. A., Ritter, G., Cregg, J. M., Old, L. J., & Batt, C. A., (2007). Cooverexpression of chaperones for enhanced secretion of a single-chain antibody fragment in Pichia pastoris. Applied Microbiology and Biotechnology, 74(2), 381–389.
15. Damasceno, L. M., Pla, I., Chang, H. J., Cohen, L., Ritter, G., Old, L. J., & Batt, C. A., (2004). An optimized fermentation process for high-level production of a single-chain Fv antibody fragment in Pichia pastoris. Protein Expression and Purification, 37(1), 18–26.
16. Das, A. J., & Deka, S. C., (2012). Fermented foods and beverages of the North-East India. International Food Research Journal, 19(2), 377.
17. De Gregorio, E., & Rappuoli, R., (2012). Vaccines for the future: learning from human immunology. Microbial Biotechnology, 5(2), 149–155.
18. Denyer, S. P., (1995). Mechanisms of action of antibacterial biocides. International Biodeterioration and Biodegradation, 36(3–4), 227–245.
19. Deo, Y. M., Mahadevan, M. D., & Fuchs, R., (1996). Practical considerations in operation and scale-up of spin-filter based bioreactors for monoclonal antibody production. Biotechnology Progress, 12(1), 57–64.
20. Engel, C. A. R., Straathof, A. J., Zijlmans, T. W., van Gulik, W. M., & van der Wielen, L. A., (2008). Fumaric acid production by fermentation. Applied Microbiology and Biotechnology, 78(3), 379–389.
21. Freitas, F., Alves, V. D., & Reis, M. A., (2011). Advances in bacterial exopolysaccharides: From production to biotechnological applications. Trends in Biotechnology, 29(8), 388–398.
22. Freyre, F. M., Vázquez, J. E., Ayala, M., Canaán-Haden, L., Bell, H., Rodríguez, I., & Gavilondo, J. V., (2000). Very high expression of an

anti-carcinoembryonic antigen single-chain Fv antibody fragment in the yeast Pichia pastoris. Journal of Biotechnology, 76(2–3), 157–163.
23. Germain, R. N., & Schwartzberg, P. L., (2011). The human condition: An immunological perspective. Nature Immunology, 12(5), 369–372.
24. Ghosh, S., Chisti, Y., & Banerjee, U. C., (2012). Production of shikimic acid. Biotechnology Advances, 30(6), 1425–1431.
25. Glevitzky, M., Dumitrel, G. A., & Popa, M., (2009). Studies regarding non-enzymatic browning of orange and apple juices during storage. Journal of Agroalimentary Processes and Technologies, 4, 511–514.
26. Gold, R., Linker, R. A., & Stangel, M., (2012). Fumaric acid and its esters: An emerging treatment for multiple sclerosis with antioxidative mechanism of action. Clinical Immunology, 142(1), 44–48.
27. Gomes, A. M., & Malcata, F. X., (1999). Bifidobacterium spp. and Lactobacillus acidophilus: biological, biochemical, technological, and therapeutical properties relevant for use as probiotics. Trends in Food Science and Technology, 10(4–5), 139–157.
28. Grage, K., Jahns, A. C., Parlane, N., Palanisamy, R., Rasiah, I. A., Atwood, J. A., & Rehm, B. H., (2009). Bacterial polyhydroxyalkanoate granules: Biogenesis, structure, and potential use as nano-/micro-beads in biotechnological and biomedical applications. Biomacromolecules, 10(4), 660–669.
29. Grigorieva, G.I., (2001). An integrated approach to biotechnology for obtaining bacterial antigens on the Brucella model. Bulletin of the Nizhny Novgorod University. NI Lobachevsky. Series: Biology, (1), 35–38.
30. Heller, K. J., (2001). Probiotic bacteria in fermented foods: Product characteristics and starter organisms. The American Journal of Clinical Nutrition, 73(2), 374–379.
31. Inan, M., Chiruvolu, V., Eskridge, K. M., Vlasuk, G. P., Dickerson, K., Brown, S., & Meagher, M. M., (1999). Optimization of temperature-glycerol-pH conditions for a fed-batch fermentation process for recombinant hookworm (Ancylostoma caninum) anticoagulant peptide (AcAP-5) production by Pichia pastoris. Enzyme and Microbial Technology, 24(7), 438–445.
32. Jayapal, K. P., Wlaschin, K. F., Hu, W., & Yap, M. G., (2007). Recombinant protein therapeutics from CHO cells-20 years and counting. Chemical Engineering Progress, 103(10), 40.

33. Jazmi, A. F., Alfiantya, P. F., Nurarifah, S. A. H., Purmitasari, E. A., Vitania, L. A., & Riawan, W., (2015). Spade leaf extract phytosome modulates Krox-20, Neuregulin-1, phospholipids, and cognitive function of traumatic brain injury model in rats. Indonesian Journal of Cancer Chemoprevention, 6(3), 105–110.
34. Josefsberg, J. O., & Buckland, B., (2012). Vaccine process technology. Biotechnology and Bioengineering, 109(6), 1443–1460.
35. Kapoor, K. K., Yadav, K. S., Singh, D. P., Mishra, M. M., & Tauro, P., (1983). Enrichment of compost by Azotobacter and phosphate solubilizing microorganisms. Agricultural Wastes, 5(3), 125–133.
36. Khan, N. T., Bibi, M., Yousuf, S., Qureshi, I. H., Al-Majid, A. M., Mesaik, M. A., & Choudhary, M. I., (2012). Synthesis of some potent immunomodulatory and anti-inflammatory metabolites by fungal transformation of anabolic steroid oxymetholone. Chemistry Central Journal, 6(1), 1–11.
37. Khanna, S., & Pardi, D. S., (2010). The growing incidence and severity of Clostridium difficile infection in inpatient and outpatient settings. Expert Review of Gastroenterology and Hepatology, 4(4), 409–416.
38. Khanna, S., & Pardi, D. S., (2012). Clostridium difficile infection: New insights into management. In: Mayo Clinic Proceedings, 87(11), 1106–1117.
39. Khatri, N. K., & Hoffmann, F., (2006). Impact of methanol concentration on secreted protein production in oxygen-limited cultures of recombinant Pichia pastoris. Biotechnology and Bioengineering, 93(5), 871–879.
40. Koff, W. C., Burton, D. R., Johnson, P. R., Walker, B. D., King, C. R., Nabel, G. J., & Plotkin, S. A., (2013). Accelerating next-generation vaccine development for global disease prevention. Science, 340 (6136), 12–20.
41. Koliasnikov, O. V., Grigorenko, V. G., Egorov, A. M., Lange, S., & Schmid, R. D., (2011). Recombinant production of horseradish peroxidase conjugates with Fab antibodies in Pichia pastoris for analytical applications. Acta Naturae (Англоязычная Версия), 3(3 (10)).
42. Lahijani, R., Hulley, G., Soriano, G., Horn, N. A., & Marquet, M., (1996). High-yield production of pBR322-derived plasmids intended for human gene therapy by employing a temperature-controllable point mutation. Human Gene Therapy, 7(16), 1971–1980.

43. Ledingham, G. A., (1953). Industrial fermentations. Annual Reviews in Microbiology, 7(1), 433–460.
44. Li, F., Vijayasankaran, N., Shen, A., Kiss, R., & Amanullah, A., (2010). Cell culture processes for monoclonal antibody production. In: MAbs, 2(5), 466–479.
45. Li, K., Mikola, M. R., Draths, K. M., Worden, R. M., & Frost, J. W., (1999). Fed-batch fermentor synthesis of 3-dehydroshikimic acid using recombinant Escherichia coli. Biotechnology and Bioengineering, 64(1), 61–73.
46. Li, Y., Chen, J., & Lun, S. Y., (2001). Biotechnological production of pyruvic acid. Applied Microbiology and Biotechnology, 57(4), 451–459.
47. Liras, A., (2008). Recombinant proteins in therapeutics: Hemophilia treatment as an example. International Archives of Medicine, 1(1), 1–4.
48. Liziyyannida, L., & Riawan, W., (2015). Anticancer effects of LBA-ST yogurt supernatant on HeLa cells via heat shock protein 27 decrease in vitro. Indonesian Journal of Cancer Chemoprevention, 6(1), 16–19.
49. Manzoni, M., & Rollini, M., (2002). Biosynthesis and biotechnological production of statins by filamentous fungi and application of these cholesterol-lowering drugs. Applied Microbiology and Biotechnology, 58(5), 555–564.
50. Markaryan, A. N., Mashko, S. V., Kukel, L. V., Lapidus, A. L., Bach, A. N., & Egorov, A. M., (1991). Construction of expression vectors for gene fusions on the model of β-galactosidase-human fibroblast β-interferon for the purpose of immunoenzyme assay. Annals of the New York Academy of Sciences, 646(1), 125–135.
51. Matsuoka, Y., & Shimizu, K., (2011). Metabolic regulation in Escherichia coli in response to culture environments via global regulators. Biotechnology Journal, 6(11), 1330–1341.
52. McGovern, P. E., Zhang, J., Tang, J., Zhang, Z., Hall, G. R., Moreau, R. A., & Wang, C., (2004). Fermented beverages of pre-and proto-historic China. Proceedings of the National Academy of Sciences, 101(51), 17593–17598.
53. Mehren, M. V., Adams, G. P., & Weiner, L. M., (2003). Monoclonal antibody therapy for cancer. Annual Review of Medicine, 54(1), 343–369.

54. Mills, S., Stanton, C., Lane, J. A., Smith, G. J., & Ross, R. P., (2019). Precision nutrition and the microbiome: Part I. Current State of the Science. Nutrients, 11 (4): 923.
55. Moharregh-Khiabani, D., Linker, R. A., Gold, R., & Stangel, M., (2009). Fumaric acid and its esters: An emerging treatment for multiple sclerosis. Current Neuropharmacology, 7(1), 60–64.
56. Palomares, L. A., Estrada-Moncada, S., & Ramírez, O. T., (2004). Production of recombinant proteins. Recombinant Gene Expression, 15–51.
57. Papagianni, M., (2007). Advances in citric acid fermentation by Aspergillus niger: Biochemical aspects, membrane transport and modeling. Biotechnology Advances, 25(3), 244–263.
58. Prasad, G. K., Ramacharyulu, P. V. R. K., Merwyn, S., Agarwal, G. S., Srivastava, A. R., Singh, B., & Vijayaraghavan, R., (2011). Photocatalytic inactivation of spores of Bacillus anthracis using titania nanomaterials. Journal of Hazardous Materials, 185(2–3), 977–982.
59. Praveen, V. K., & Savitha, J., (2012). Solid-state fermentation: An effective method for lovastatin production by fungi over submerged fermentation. E3 Journal of Biotech. Pharma Res., 3(2), 15–21.
60. Rahman, M., & Shimizu, K., (2008). Altered acetate metabolism and biomass production in several Escherichia coli mutants lacking rpoS-dependent metabolic pathway genes. Molecular BioSystems, 4(2), 160–169.
61. Rahman, M., (2013). Medical applications of fermentation technology. In: Advanced Materials Research, 810, 127–157.
62. Rahman, M., Hasan, M. R., & Shimizu, K., (2008). Growth phase-dependent changes in the expression of global regulatory genes and associated metabolic pathways in Escherichia coli. Biotechnology Letters, 30(5), 853–860.
63. Rappa, G., Lorico, A., & Sartorelli, A. C., (1993). Reversal of etoposide resistance in non-P-glycoprotein expressing multidrug-resistant tumor cell lines by novobiocin. Cancer Research, 53(22), 5487–5493.
64. Rehm, B. H., (2010). Bacterial polymers: biosynthesis, modifications, and applications. Nature Reviews Microbiology, 8(8), 578–592.
65. Rodrigues, M. E., Costa, A. R., Henriques, M., Azeredo, J., & Oliveira, R., (2010). Technological progresses in monoclonal antibody production systems. Biotechnology Progress, 26(2), 332–351.

66. Rohlke, F., & Stollman, N., (2012). Fecal microbiota transplantation in relapsing Clostridium difficile infection. Therapeutic Advances in Gastroenterology, 5(6), 403–420.
67. Samuel, B. P., Crumb, T. L., & Duba, M. M., (2014). What nurses need to know about fecal microbiota transplantation: Education, assessment, and care for children and young adults. Journal of Pediatric Nursing, 29(4), 354–361.
68. Saroj, S., Kumar, K., Prasad, M., & Singh, R. P., (2014). Differential expression of peroxidase and ABC transporter as the key regulatory components for degradation of azo dyes by Penicillium oxalicum SAR-3. Functional and Integrative Genomics, 14(4), 631–642.
69. Schmid, J., Sieber, V., & Rehm, B., (2015). Bacterial exopolysaccharides: Biosynthesis pathways and engineering strategies. Frontiers in Microbiology, 6, 496.
70. Seow, Y., & Wood, M. J., (2009). Biological gene delivery vehicles: Beyond viral vectors. Molecular Therapy, 17(5), 767–777.
71. Shiloach, J., & Fass, R., (2005). Growing E. coli to high cell density: A historical perspective on method development. Biotechnology Advances, 23(5), 345–357.
72. Shimizu, K., (2004). Metabolic flux analysis based on 13 C-labeling experiments and integration of the information with gene and protein expression patterns. Recent Progress of Biochemical and Biomedical Engineering in Japan Ii, 1–49.
73. Shukla, A. A., & Thömmes, J., (2010). Recent advances in large-scale production of monoclonal antibodies and related proteins. Trends in Biotechnology, 28(5), 253–261.
74. Singh, A. K., Athmaram, T. N., Shrivastava, S., Merwyn, S., Agarwal, G. S., & Gopalan, N., (2013). Fermentation and downstream process for high yield production of Plasmodium falciparum recombinant HRP II protein and its application in diagnosis. Journal of Industrial Microbiology and Biotechnology, 40(7), 687–695.
75. Song, H., (1991). The challenge of AIDS in China. AIDS Patient Care, 5(3), 146–151.
76. Soomro, F. R., Pathan, G. M., Bajaj, D. R., & Bhatti, N. S., (2012). Leprosy in Larkano region; An analysis of 102 cases from 2001–2011 at leprosy center Larkano, Sindh, Pakistan. Journal of Pakistan Association of Dermatologists, 22(2).

77. Stanbury, P. F., Whitaker, A., & Hall, S. J., (1995). Microbial growth kinetics. Principles of Fermentation Technology (pp. 13–34). Butterworth Heinemann, Oxford, United Kingdom.
78. Subramaniyam, R., & Vimala, R., (2012). Solid-state and submerged fermentation for the production of bioactive substances: A comparative study. Int. J. Sci. Nat., 3(3), 480–486.
79. Suno, M., & Nagaoka, A., (1989). Inhibition of lipid peroxidation by idebenone in brain mitochondria in the presence of succinate. Archives of Gerontology and Geriatrics, 8(3), 291–297.
80. Tachibana, H., Takekoshi, M., Cheng, X. J., Nakata, Y., Takeuchi, T., & Ihara, S., (2004). Bacterial expression of a human monoclonal antibody-alkaline phosphatase conjugate specific for Entamoeba histolytica. Clinical and Vaccine Immunology, 11(1), 216–218.
81. Tang, Y. J., Ohashi, R., & Hamel, J. F. P., (2007). Perfusion culture of hybridoma cells for hyperproduction of IgG2a monoclonal antibody in a wave bioreactor-perfusion culture system. Biotechnology Progress, 23(1), 255–264.
82. Tian, C., Zhao, J., Zhang, J., Chu, S., Dang, Z., Lin, Z., & Xing, B., (2017). Enhanced removal of roxarsone by Fe 3 O 4@ 3D graphene nanocomposites: Synergistic adsorption and mechanism. Environmental Science: Nano, 4(11), 2134–2143.
83. Vaishnavi, C., (2014). Fecal microbiota transplantation for management of Clostridium difficile infection. Indian Journal of Gastroenterology, 33(4), 301–307.
84. Vicente, T., Mota, J. P., Peixoto, C., Alves, P. M., & Carrondo, M. J., (2011). Rational design and optimization of downstream processes of virus particles for biopharmaceutical applications: Current advances. Biotechnology Advances, 29(6), 869–878.
85. Vinderola, C. G., Mocchiutti, P., & Reinheimer, J. A., (2002). Interactions among lactic acid starter and probiotic bacteria used for fermented dairy products. Journal of Dairy Science, 85(4), 721–729.
86. Vinderola, C. G., Prosello, W., Ghiberto, D., & Reinheimer, J. A., (2000). Viability of probiotic (Bifidobacterium, Lactobacillus acidophilus and Lactobacillus casei) and nonprobiotic microflora in Argentinian Fresco cheese. Journal of Dairy Science, 83(9), 1905–1911.
87. Wang, C. Y., & Walfield, A. M., (2005). Site-specific peptide vaccines for immunotherapy and immunization against chronic diseases, cancer,

infectious diseases, and for veterinary applications. Vaccine, 23(17–18), 2049–2056.

88. Wang, C. Y., Finstad, C. L., Walfield, A. M., Sia, C., Sokoll, K. K., Chang, T. Y., & Windisch, M., (2007). Site-specific UBITh® amyloid-β vaccine for immunotherapy of Alzheimer's disease. Vaccine, 25(16), 3041–3052.
89. Wang, C. Y., Shen, M., Tam, G., De Fang, X., Ye, J., Shen, F., & Hanson, C. V., (2002). Synthetic AIDS vaccine by targeting HIV receptor. Vaccine, 21(1–2), 89–97.
90. Wang, Z., Zhuge, J., Fang, H., & Prior, B. A., (2001). Glycerol production by microbial fermentation: A review. Biotechnology Advances, 19(3), 201–223.
91. Wu, N., & Ataai, M. M., (2000). Production of viral vectors for gene therapy applications. Current Opinion in Biotechnology, 11(2), 205–208.
92. Yastremskaya, K. Y., Markov, E. Y., Nikolaev, V. B., Mikhailov, L. M., Barannikova, N. L., Tokareva, L. E., & Balakhonov, S., (2014). Approaches to the development of erythrocyte diagnosticum for the detection of specific antibodies against brucella in the L-form. Journal of Infectious Pathology, 21 (1–2), 113–114.
93. Zhinan, X., & Peilin, C., (1999). Enhanced production of avermectin B1a by medium optimization and glucose feeding with Streptomyces avermilitis. Bioprocess Engineering, 20(1), 67–71.

INDEX

C

D

E

F

G

H

I

L

R

S

T

U

V

W

X

Y

Z